Ordália Dias da Silva Guilherme
Adão Francisco
Geovanny Guilherme

The Sustainable Development Master Plan for Porto Nacional - TO

Ordália Dias da Silva Guilherme
Adão Francisco
Geovanny Guilherme

The Sustainable Development Master Plan for Porto Nacional - TO

The management of urban environmental instruments

ScienciaScripts

Imprint

Any brand names and product names mentioned in this book are subject to trademark, brand or patent protection and are trademarks or registered trademarks of their respective holders. The use of brand names, product names, common names, trade names, product descriptions etc. even without a particular marking in this work is in no way to be construed to mean that such names may be regarded as unrestricted in respect of trademark and brand protection legislation and could thus be used by anyone.

Cover image: www.ingimage.com

This book is a translation from the original published under ISBN 978-613-9-62739-4.

Publisher:
Sciencia Scripts
is a trademark of
Dodo Books Indian Ocean Ltd. and OmniScriptum S.R.L publishing group

120 High Road, East Finchley, London, N2 9ED, United Kingdom
Str. Armeneasca 28/1, office 1, Chisinau MD-2012, Republic of Moldova, Europe
Printed at: see last page
ISBN: 978-620-7-74978-2

SUMMARY

DEDICATORY

I dedicate this work, first and foremost, to God, the supernatural force, the master of life, who knows all my dilemmas and who, through faith, has given me the resilience, wisdom and motivation to carry on. For the life he gives me, accompanied by the health and vigor to fight for my ideals. I depend and will always depend on my God!

And secondly, she is a utopian dreamer, who dreams of a fairer, more equal and fraternal world, and whose instruments of struggle are teaching, research and politics. She considers her dream to be difficult to achieve and that's why she works hard to overcome this gap and considers herself a happy person because the fruit of her work is beautiful. Consider this dissertation part of that fruit, Professor Dr. Adao Francisco de Oliveira.

ACKNOWLEDGMENTS

To my God, the supernatural force that I have sought through faith and am victorious over.

To my husband Nelcione, who has been a true friend and companion throughout this journey, without losing faith in me.

To my children: Geovanny and Sâmella, my heart outside my body.

To the coordinator of the Master's Program, who is more than a teacher, a true friend, committed to the development of education in Porto Nacional and the region, Dr. Elizeu Lira.

To my advisor, Professor Dr. Adao Francisco de Oliveira, for his attention, encouragement, dedication, in short, for everything he has done for me directly and indirectly.

To Germana Coriolano, who was always very friendly and patiently welcomed us into her home for some guidance, along with her partner, my advisor.

To my professors in the Master's Program: Dr. Ariovaldo Umbelino, Dr. Elizeu Ribeiro Lira, Dr. Roberto de Souza Santos, Dr. Kelly Cristine Fernandes de Oliveira Bessa, Dr. Fernando Morais, Dr. José Ramiro Lamadrid Maron, Dr. Lucas Barbosa e Souza and Dr. Rosane Balsan.

To Professor Dr. Lucas Barbosa, especially, who has been with me since the presentation of the Project, in the qualification of this dissertation and in the examining board. All his contributions were of great importance.

To the staff at the Master's office, Raimunda and Ronaldo, who were always helpful.

To my classmates for the friendship built up over the years: Claudia Fernanda Pimentel, Claudiomar da Cruz Martins, Diógenes Alencar Bolwerk, Edna de Jesus, Guilherme Pereira de Carvalho, Joeslan Rocha, Junio Batista do Nascimento, Maria de Jesus Coelho Abreu and Nùbia Nogueira do Nascimento.

To the secretaries of the municipal legislature of Porto Nacional: Maria da Conceiçao (Ceiça) and Rayce Cristina, for their kindness and patience in offering access to the legislature's collection.

To my friend Professor Sebastiao, for mediating in the town hall.

To Aurenice, the director of the Housing and Environment Department, and Sonaira, an environmental analyst at the city hall, who dedicatedly gave me their attention and made available the reports on the activities carried out by the Housing and Environment Department.

To Councillor Ronivom for mediating with the Municipal Planning Department.

To the staff of the Porto Nacional City Hall for the information provided.

Finally, I would like to thank all my family, especially my mother, my mother-in-law and my brothers, who, even though they didn't understand much of what was going on, tried to understand and accept me the way I am.

SUMMARY

The process of planning and managing urban space has gained prominence in debates related to models of occupation and reproduction of space. The disorderly condition of this occupation, observed in countless Brazilian cities, has contributed to the emergence and worsening of urban environmental problems. Based on this framework, this research aims to analyze Porto Nacional's Sustainable Development Master Plan (PDDSPN) as an environmental management instrument and its application in the light of the guidelines set out in the City Statute between 2002 and 2014. The research focuses in particular on evaluating environmental zoning as a planning and management instrument in the PDDSPN. In this sense, it seeks to understand the following situations: what are the complementary laws that establish these study instruments and what do they recommend; what are the plans, programs and projects (actions) that the city government applies and that correspond to the city's urban environmental policy; what was the budget execution for these policies; and who was the public directly and indirectly affected by them. To this end, the methodology is based on different techniques, such as documentary analysis and fieldwork. After a theoretical and conceptual discussion of the planetary urbanization process, focusing on Brazil and Tocantins from a more holistic perspective, the way is opened to explore the political and legal instruments of urban planning in Brazil, from the perspective of the federative relationship typical of the Brazilian republican regime, considering the role of social movements in the construction of the urban reform chapter in the 1988 Constitution, as well as the drafting of the City Statute and, subsequently, the Master Plans with their attributions. The aim is not only to understand the meaning of the PDDSPN, but also to extract the points it makes in defining the guidelines that focus on the urban environmental issue. The aim is also to assess the effectiveness of the Macrozones in preserving conservation units, which are considered fundamental elements in guaranteeing urban amenity. In this sense, the phenomenon of urbanization makes sense in the light of the efforts to explain it in two fields of knowledge that stand out: Urbanism, as knowledge for aesthetic-structural intervention; and Urban Planning, as knowledge of management and political intervention in urban space.

Keywords: City Statute. Master Plan. Urban Environmental Management. Macrozoning. Porto Nacional-TO.

INTRODUCTION

In July 2001, Law No. 10.257, known as the City Statute, was sanctioned, regulating articles 182 and 183 of the Federal Constitution, which set out important basic principles to guide urban policy actions. This law was the result of historic popular demands regarding the right of all citizens to the city. The main objective of this law is to establish norms that regulate the use of urban land and discipline the means of urban policy action, such as improving safety, people's well-being and environmental balance.

The City Statute has 58 (fifty-eight) articles, divided into five chapters, which deal with: 1) the General Guidelines; 2) the Urban Policy Instruments; 3) the Master Plan; 4) the Democratic Management of the City; and 5) the General Provisions.

It is in the instruments of urban policy, found in Chapter II, between Articles 4 and 38 of the City Statute, which concern municipal planning, that the fundamental aspect of this study's investigation is located: the instruments of urban environmental management. In terms of urban policy, the following instruments stand out:

I.   MUNICIPAL PLANNING: environmental zoning; sectoral plans, programs and/or projects (policies);

II.   DOSINSTITUTOS TRIBUTARIOS E FINANCEIROS: IPTU progressivo;

improvement contribution; tax and financial incentives;

III.   LEGAL AND POLITICAL INSTITUTES: expropriation; tombamento;

establishment of conservation units; establishment of ZEIs; usucapiao; onerous grants; pre-emption rights; consortium urban operations, etc;

IV.   EIA/EIV- Neighborhood impact.

Also provided for in article 182 of the 1988 Federal Constitution and regulated in articles 39 to 42 of the City Statute, the Master Plan is presented as a fundamental instrument for urban planning, as it defines the policy for urban development and expansion, establishing a model compatible with the protection of natural resources and in defense of the population's well-being.

Although the rules contained in the City Statute are more directly linked to the field of urban planning law and not environmental law, they have clear repercussions for the protection not only of the built environment, but also of the natural environment.

In this sense, the main objective of the Master Plan, according to Santos Jùnior and Montandon (2011, p. 14), is to define the social function of the city and urban property in order to guarantee "access to urbanized and regularized land for all social segments, to guarantee the right to housing and urban services for all citizens, as well as to implement a policy of democratic and participatory management [...]".

According to Brasil (2002), the City Statute makes it compulsory for municipalities to have a Master Plan law: (i) with more than twenty thousand inhabitants, (ii) which are part of metropolitan regions and urban agglomerations, (iii) where the municipal government intends to use the instruments provided for in paragraph 4° of article 182 of the Federal Constitution. 182 of the Federal Constitution, (iv) in areas of special tourist interest, (v) in the area of influence of projects or activities with a significant regional or national environmental impact and (added by Law No. 12.608 of 2012) (vi) included in the national register of municipalities with areas susceptible to the occurrence of major landslides, flash floods or related geological or hydrological processes.

The municipality of Porto Nacional falls directly into two of these situations: the first, because it has more than 20,000 inhabitants; and the fifth, because it is located in an area with significant environmental impact, resulting from the damming of the Tocantins River for the construction of the Luis Eduardo Magalhaes Hydroelectric Power Station. It can also be included in the fourth situation, because it is an area of special tourist interest, related to its historical, architectural and cultural heritage. These situations justify the existence of its Master Plan.

As far as the legal-institutional situation is concerned, Porto Nacional is governed by the Municipal Organic Law, approved on April 4, 1990, and in 2006, by Municipal Complementary Law No. 05/2006, when the Porto Nacional Sustainable Development Master Plan (PDDSPN) came into force. This document, which has 95 articles and 26 pages, had the participation of various agents involved in its preparation, including: the public authorities (the municipal executive, some elements of the state and federal executives, the Municipal Legislature and the State Public Prosecutor's Office), economic agents and the community. These players established the following objectives in the document, as guiding principles for the municipality's development: strengthening its agricultural vocation and expanding the agro-industrial park; consolidating it as a regional fruit and vegetable hub; structuring it as a center of excellence in education; valuing its historical, cultural and tourist heritage; encouraging the fight against hunger and social exclusion (PORTO NACIONAL, 2006).

The PDDSPN is the starting point for this research, based on the environmental planning instruments highlighted in articles 25 and 26. The former establishes the planning of the municipal territory by means of macro-zoning, the structuring of vehicular and pedestrian circulation routes and the preservation of historical and cultural heritage. Of the three elements presented, the establishment of the macro-zoning and the management devices created for its implementation and effectiveness are the main focus of this research.

It is important to highlight what the macro-zoning instrument is all about. According to Article 26 of the PDDSPN,

Macrozoning is understood to be the division of the municipal territory into integrated areas, called macrozones, with the aim of promoting its planning, as well as the planning and proper implementation of the strategic lines and action programs defined by Porto Nacional's Sustainable Development Master Plan (PORTO NACIONAL, 2005).

In Porto Nacional, the macro-zoning was established using the following categories:

I - housing

II   - commercial and services

III   - industrial

IV   - landscape-environmental

V   - special

We are interested in focusing on the landscape-environment category mentioned in Article 7 of the aforementioned plan, as it deals with areas that require special treatment due to their natural characteristics and their landscape, environmental and tourist potential.

Before the PDDSPN, Municipal Law no. 1782, of December 27, 2003, established the Territorial Macrozoning in the Municipality of Porto Nacional as a municipal planning instrument. Article 27 of the PDDSPN establishes the macro-zoning in three large distinct areas, seeking to articulate environmental, economic, cultural and social variables in the following macro-zones: i. Urban Macrozone (MU); ii. Environmental Protection Macrozone (MA); and iii. Rural Macrozone (MR).

The area spatially delimited for this study is included in the first two macro-zones: the Urban Macrozone (MU) and the Environmental Protection Macrozone (MA), as it is understood that a healthy environment is a right to life with dignity. And the interdependence between the urban environment and equity with the natural environment makes them more amenable to the most varied problems such as health, education, solid waste, human settlements, climate, protection of people, water resources and protection of ecosystems.

In this sense, it is necessary to understand the concept of the environment and its relationship with urban environmental management, as it is the basis for discussing the environmental guidelines and instruments of the City Statute included in Porto Nacional's Sustainable Development Master Plan.

The environment can be conceptualized in the light of Federal Law no. 6.938, of 31/8/1981, article 3 and item I, which establishes the National Environmental Policy: "a set of conditions, laws, influences and interactions of a physical, chemical and biological order, which allows, shelters and governs life in all its forms".

Prieto (2006) classifies the environment into natural elements (water, air, soil, flora, fauna), responsible for the environmental balance between living beings and the environment in which they live; and the built or artificial, formed by the set of buildings and equipment that make up the urban space.

Prieto (2006, p. 4) reports that the clash is actually between environmental and social issues in the urban environment and presents three analytical perspectives for this understanding. The first is the association of the environmental only with the natural, when we know that it also includes social relationships and dynamics. The second is that the environment, as a result of the relationship between the natural and the social, must be analyzed from a temporal perspective, i.e. while nature counts time in millennia and centuries (geological eras), society counts in years, days and hours. The third perspective presents the clash between environmental and social issues as social representations that are constructed about nature and the city in contemporary society, precisely that which marks the schism between the natural and the urban, between the world of the countryside and the world of commerce, which is the relevant clash for the research in focus.

The author adds that urban and environmental management should be understood as the set of technical, administrative, legal and regulatory activities dedicated to the management of a city, in which the improvement or conservation of environmental quality, both in the intra-urban space and in its area of influence, represents a determining objective. According to Prieto (2006), cities occupy no more than 5% of the world's land area, yet they alter rivers, woodlands and cultivated fields, forests and the atmosphere. And these changes are mainly due to the use and appropriation of soil, water and air by various human agents.

Rodrigues (1998) states that the environmental issue should be understood as a product of society's intervention in nature. It concerns not only problems related to nature, but also issues arising from social action, i.e. society plays a full part in the construction of the urban and environmental environment, intervening in a beneficial way or not.

A brief discussion of "Sustainable Development" is also worth mentioning here, as it is a complementary part of Porto Nacional's Master Plan and can only be understood from a theoretical perspective. Thus, Seiffert (2011) states that environmental concern with natural resources only appeared in the 20th century, around the 1960s, with the intensive use of natural resources and the increase in population and consumption, factors of great relevance in environmental management. But the concept of Sustainable Development really emerged in the 1970s with the Stockholm Conference in Sweden, where the consequences of the capitalist economy on the environment were discussed. As a result, the Declaration on the Human Environment and a World Plan of Action emerged, with the aim of guiding the rational use of natural resources and improving the human environment.

In the 1980s, as a result of the formalization of Environmental Impact Studies and Impact Reports on the Environment - EIA-RIMA, the First World Conservation Strategy in collaboration with the United Nations Environment Programme - UNEP and the World Wildlife Fund - WWF, the concept of "Sustainable Development" emerged for the first time, presenting a long-term plan to conserve the planet's biological resources. The World Commission on Environment and Development (WCED), set up by the UN in 1983, based on reports, pointed to inequality between countries and poverty as the main causes of environmental problems. Since then, Seiffert (2011, p. 14-15) has defined two key concepts,

[...] that of needs, especially the essential needs of the world's poor, which must be given top priority, and a sense of the limitations that the state of technology and social organization imposes on the environment, preventing it from meeting the needs of future generations.

The author states that sustainable development can only be achieved by balancing the imperatives of the environmental, social and economic spheres. However, Seiffert (2011, p. 24) emphasizes that

Although it is a widely used concept, there is no single vision of what sustainable development is. For some, achieving sustainable development means achieving continued economic growth through more rational management of natural resources and the use of more efficient and less polluting technologies. For others, sustainable development is above all a social and political project aimed at eradicating poverty, raising the quality of life and satisfying the basic needs of humanity, which offers the principles and guidelines for the harmonious development of society, taking into account the appropriation and sustainable transformation of environmental resources.

Rodrigues (2005) severely criticizes sustainable development and treats it as a myth of the environmental issue, which has been elevated to the top of the political agenda. Thus, all issues and problems refer to the environment as a common good and a necessity for future generations. He believes that accepting sustainable development in relation to the environment provides a wide range of alternatives due to the vagueness of the term itself. It also states that legitimizing sustainable development with definitions of social, political, economic, territorial, ecological and spatial sustainability contradicts other definitions. For example, economic sustainability contradicts the idea of social sustainability, because the term "sustainable development" is not a concept, but an idea that aims to find solutions to problems of depletion and pollution of natural resources in the future. A generic idea that abstracts reality and hides complexity, because we use "natural wealth" as a counterpoint to "natural resources", the latter characterizing the elements of nature as merchandise. Reflexivity of the commodity mode of production creates a thick smokescreen over the appropriation of territories and the existence of social classes, making critical analysis difficult.

In this sense, the purpose of this dissertation is to analyze the Porto Nacional-Tocantins Sustainable

Development Master Plan, as an environmental management instrument, and its application in the light of the guidelines contained in the City Statute. The research was carried out by evaluating the application of the main urban planning instruments included in the City Statute, which are present in the Sustainable Development Master Plan of the municipality of Porto Nacional in the state of Tocantins.

Francisco de Oliveira (2011) evaluates Porto Nacional's Sustainable Development Master Plan as a document that sought to appropriate most of the urban planning instruments provided for in the City Statute and defined them in accordance with its conceptualization, understanding that, however, the plan did not define how they would be applied, thus requiring subsequent regulations. It considers popular participation and the involvement of all social actors to be important in the construction of a collective territorial planning project, which must be guaranteed through the creation of its Municipal Planning and Development Council and the establishment of the City Conference as an organ belonging to the Master Plan Management structure.

According to the same author, the technical documents guiding the drafting of the Law creating the Master Plan are well prepared and maintain a good level of coherence with the concept of sustainability. However, the procedures need to be better detailed so that the instruments included in the PDDSPN can be complied with.

In this circumstance, we can infer that the PDDSPN has sufficient elements for an environmental applicability that guarantees the sustainable development of the city, as well as the well-being of the population of the municipality.

It is hoped that this analysis will contribute to an assessment of Porto Nacional's territorial planning and management, as well as to guiding the implementation of urban policy instruments, with a view to urban growth that ensures a social balance, through the fair distribution of the benefits and burdens derived from the urbanization process. Under the 1988 Federal Constitution, art. 23, municipalities are also responsible for protecting the environment: conserving public heritage; protecting documents, works and other goods of value (historical, artistic and cultural), monuments, notable natural landscapes and archaeological sites; providing the means of access to culture, education and science; protecting the environment and combating pollution; preserving forests, fauna and flora; and registering, monitoring and supervising the granting of rights to research and exploit water and mineral resources in their territories (BRASIL, 1988).

Given this premise, the challenge is to study the theoretical and conceptual foundations, as well as the legal frameworks of Brazilian environmental and urban policies, in order to gather information and answer the essential question raised here: **how does Porto Nacional's Sustainable Development Master Plan present itself as an environmental management instrument and its**

application in the light of the guidelines contained in the city's statute?

It is based on the premise that the city of Porto Nacional arose without any urban planning, promoting the destruction of natural resources in an indiscriminate and perceptible way. The central question this dissertation seeks to answer is whether Porto Nacional's Sustainable Development Master Plan has established effective urban management instruments to minimize these environmental impacts.

This question leads to other questions aimed at understanding how the Plan is applied: what are the complementary laws that establish these planning and management instruments? What are the plans, programs and projects (actions) that the city government is implementing that have an impact on the implementation of these instruments? What was the budget execution? And which audiences have been reached? What are the limitations and potentialities of the planning process for municipal management?

Thus, the specific objectives that emerged from the main question are:

a)   to review the conceptual basis of urbanization, urbanism and urban planning in Brazil and Tocantins;

b)   understanding the political and legal instruments of urban planning in Brazil;

c)   to analyze the environmental implications of the environmental instruments of the City Statute and their application in relation to Porto Nacional's Sustainable Development Master Plan;

d)   analyze the interface/dialogue between environmental and urban policy instruments in the municipality of Porto Nacional to achieve the construction of urban environmental management in the face of environmental zoning.

Although it uses quantitative data, this investigation has a qualitative emphasis, which is essentially exploratory in nature, allowing for greater discussion of the subject under investigation.

For Silveira and Córdova (2009), researchers who use qualitative methods seek to explain why things happen, expressing what should be done, because qualitative research has: objectification of the phenomenon; hierarchization of the actions of describing, understanding, explaining, precision of the relationships between the global and the local in a given phenomenon; observance of the differences between the social world and the natural world; respect for the interactive nature between the objectives sought by the researchers, their theoretical orientations and their empirical data; search for the most reliable results possible; opposition to the assumption that defends a single research model for all sciences.

Minayo (2003) points out that qualitative research works with the universe of meanings, motives, aspirations, beliefs, values and attitudes, which corresponds to a deeper space of relationships, processes and phenomena that cannot be reduced to the operationalization of variables. According to this author, qualitative research is the path of thought to be followed and occupies a central place in theory. Although it deals with techniques used to construct reality, it also requires a stance on the part of the researcher towards the phenomena they are studying.

Similarly, Gil (1999) explains that this type of study aims to provide the researcher with greater knowledge about the subject, so that they can formulate more precise problems or create hypotheses that can be investigated in subsequent studies. This strategy was used to clarify the problem and explain it in detail.

The conceptual analysis of urbanization, urbanism and urban planning in Brazil and Tocantins was based on the authors Spósito (1988), Hauser and Schnore (1975), Castells (2009), Santos (2009), Lira (1995), Lefebvre (2009), Souza (2005; 2010), Santos Jûnior (2011) and others. The analysis of the theme through the authors cited aims to identify the concepts and theories related to Urban Planning, with a historical reading of this process and its relationship with the elaboration of Master Plans. In the construction of the other chapters, it was necessary to apply techniques such as: documentary analysis, to find the results of the evaluation of the application of public environmental policies, planning and budget execution for environmental policies, plans, programs and projects carried out; and field research, to detect concrete actions that imply the use and conservation of natural resources.

From this perspective, the first chapter seeks to situate the municipality: its location, its historical settlement process and its socio-economic characteristics. The second chapter aims to present the meaning of the theoretical and conceptual discussion on the process of global urbanization, with a focus on Brazil and Tocantins, from a perspective of totality. The intention is to situate the phenomenon in Tocantins from its dialectical interaction with the national and global movement, exploring the main explanatory theoretical nuances and the most articulated concepts to this end. In this sense, the phenomenon of urbanization makes sense in the light of the efforts to explain it in two fields of knowledge that are at the forefront: Urbanism, as knowledge for aesthetic-structural intervention; and Urban Planning, as knowledge of management and political intervention in urban space.

This theoretical and conceptual understanding paves the way for exploring the political and legal instruments of urban planning in Brazil, from the perspective of the federative relationship typical of the Brazilian republican regime. Thus, the political and legal application of urban planning in Tocantins is based on the normative framework drawn up by the Union. On the other hand, this application must obey the specific territorial dynamics of the federative unit. The role of social movements in the construction of the urban reform chapter in the 1988 Constitution is considered, as well as the drafting of the City Statute and, subsequently, the Master Plans with their attributions. This is the discussion that takes place in the third chapter.

Reducing the scale and broadening the focus, the fourth chapter aims to deal with the main instrument of Urban Planning and Management advocated in the City Statute (the main regulatory framework for Urban Planning in Brazil): the Master Plan. The aim is not only to understand the meaning of Porto Nacional's Sustainable Development Master Plan (PDDSPN), but also to extract its notes on the definition of guidelines that focus on the urban environmental issue. It also deals with the urban environmental issue in Porto Nacional from the point of view of a specific planning instrument in the Master Plan: the establishment of Macrozones. The idea is to analyze the way in which these were included in the plan and the policies (programs, projects, actions) that resulted from them in order to consider them in the day-to-day urban experience of Porto Nacional, especially in the articulation between the built urban environment, which is eminently artificial, and the natural environment maintained by the conservation units. The aim was to evaluate the effectiveness of environmental public policies, which are considered fundamental elements in guaranteeing environmental quality.

1. HISTORICAL CONTEXT OF THE MUNICIPALITY OF PORTO NACIONAL-TO

With a population of 49,146 (IBGE, 2010), Porto Nacional is a historic city located on the right bank of the Tocantins River, between parallels 10° S and 11° S and meridians 48°W and 49°W, as shown in Map 1.

Map 1 - Location of Porto Nacional

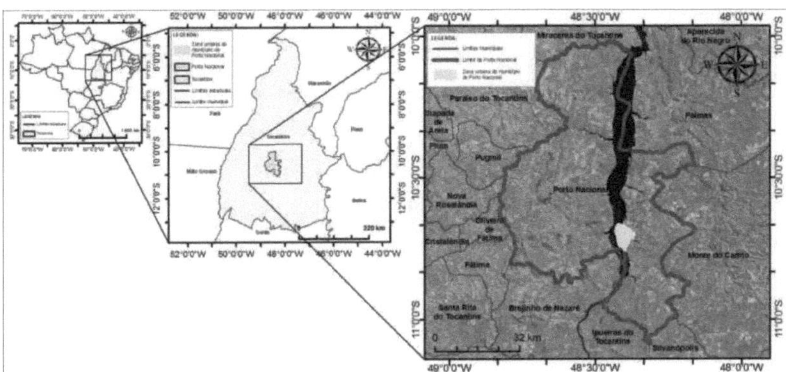

Source: MORAES, Erton Inacio M. de. (2014).

The state of Tocantins has 139 (one hundred and thirty-nine) municipalities. According to IBGE population data, Porto Nacional is the fourth largest city in the state, behind Palmas, Araguaina and Gurupi.

According to the state's regional planning, organized by SEPLAN's Ecological and Economic Zoning Directorate (2013), in terms of economic aspects, Porto Nacional's municipal GDP increased by 37.82% in 2010 compared to the previous year, representing 3.89% of the state's GDP, making it the fourth largest Gross Domestic Product in Tocantins, with the industrial sector growing the most. In 2010, the service sector accounted for 49.55% of the value added, industry 39.37% and agriculture 11.08%. The highlight of the service sector is Public Administration, which accounted for 45.6%. In the industrial sector, the activity with the largest share this year was construction, which accounted for 52.2% of the sector. In Agriculture, the production of soybeans, cassava and sugar cane and the raising of cattle, poultry and pigs stand out.

Porto Nacional has a number of projects of economic importance on the regional scene: (i) the Sao Joao Irrigation Project, located in the municipality of Porto Nacional, 25 km from Palmas-Tocantins, which began in 2001 and includes the implementation of infrastructure for irrigation of an area of 3,654 ha for growing fruit; (ii) Granol, a Brazilian company dedicated to the production and marketing of grain, bran, vegetable oils and biodiesel for the domestic and foreign markets. In Tocantins, the company has a warehouse and a biodiesel plant in Porto Nacional. In Brazil, the company has five industrial complexes, 41 regional grain purchasing and storage facilities, three biodiesel plants, a lecithin factory, a sea terminal and a river terminal, as well as its head office in Sao Paulo.

With regard to the social development of the municipality, in 2012, Atlas Brasil (2010) reported that UNDP Brazil, IPEA and the Joao Pinheiro Foundation took on the challenge of adapting the global HDI methodology to calculate the Municipal HDI (MHDI) of the 5,565 Brazilian municipalities using data from the 2010 Demographic Census. The MHDI was also recalculated, based on the methodology adopted, for the years 1991 and 2000, by means of a thorough reconciliation of municipal areas between 1991, 2000 and 2010, to take into account the administrative divisions that took place during the period and allow for temporal and spatial comparability between municipalities. The Brazilian MHDI follows the same three dimensions as the Global MHDI - longevity, education and income - but goes further: it adapts the global methodology to the Brazilian context and to the availability of national indicators. Thus, the MHDI - includes its three components, MHDI Longevity, MHDI Education and MHDI Income. The MHDI is accompanied by more than 180 socio-economic indicators, which support the analysis of the MHDI and broaden the understanding of phenomena and dynamics related to municipal development.

According to Atlas Brasil (2010), Porto Nacional's MHDI is as shown in Table 1.

Table 1 - Porto Nacional's MHDI

YEAR	INCOME	LONGEVITY	EDUCATION	MHDI	POPULATION	GROWTH. POP. %
1991	0,588	0,640	0,203	0,424	43.325	-
2000	0,619	0,708	0,406	0,562	44.991	3,7
2010	0,699	0,826	0,701	0,740	49.146	8,4

Source: UNDP/IPEA/FJP, 2013. Adapted by the author.

Porto Nacional's Municipal Human Development Index (MHDI) places the municipality in the High Human Development range (MHDI between 0.700 and 0.799). Between 2000 and 2010, the dimension that grew the most was Education, with an increase of 0.295, followed by Longevity and Income. Also between 1991 and 2000, the dimension that grew the most was Education, with an increase of 0.203, followed by Longevity and Income, as shown in Table 1.

Graph 1 shows the growth in the MHDI of all the Brazilian municipalities and shows that Porto

Nacional has evolved significantly, surpassing the state average, which is in line with the national average, even surpassing it at the end of 2010.

Graph 1 - Evolution of the MHDI - Porto Nacional-TO

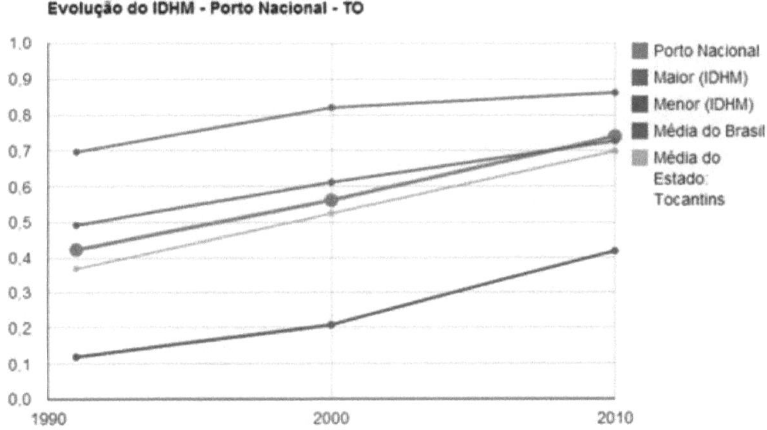

Source: UNDP/IPEA/FJP (2013).

With regard to the history of Porto Nacional, SEPLAN (2013) and IBGE (2010) state that it is one of the oldest municipalities in the state. The history of its settlement stems from gold mining, which began in 1722 in the Province of Goiás. This brought many miners to the northern region of Goiás and was responsible for most of the small towns that were established there. These miners, drovers, peddlers and travelers were already crossing in the boats of the Portuguese Félix Camôa, in the place where the historic center of Porto Nacional is today, when, in 1791, Corporal Thomaz de Souza Villa Real, who was checking out the possibility of navigation and the establishment of a south-north trade route, set up a military detachment in the region. The privileged location between two important mining towns, Pontal and Carmo, gave rise to Porto Real, which developed as a result of trade and navigation. Lira (2011, p. 122) reports that, in 1725, "the Tocantins River became an important transportation route, integrating the interior of the Northern Province of Goiás with the province of Grao-Pará, with a route that connected Belém do Parà to the Goiás city of Porto Nacional". This situation is shown in Photo 1.

Photo 1 - Sailing on the Tocantins River from Belém-PA.

Source: Author's personal archive.

Also according to the IBGE (2010), in 1831, the town of Porto Real was elevated to the category of town, changing its name to Porto Imperial. The main factors that contributed to its elevation to the seat of the municipality are: the increase in navigation on the Tocantins and trade with Belém do Parà; the decline of mining in neighboring towns, such as Pontal; and the development of cattle breeding.

When it was elevated to city status by Provincial Resolution 333 of July 13, 1861, Porto Imperial was an important commercial emporium, with many merchants, intense river trade with the north and 4,313 inhabitants. With the Proclamation of the Republic in 1889, the city was renamed Porto Nacional.

In 1886, the first Dominican priests arrived in the city and, in 1904, the Dominican sisters intensified their work in the field of education, making Porto Nacional a benchmark in the area, which attracted students from various municipalities. Photos 2 and 3 show the Dominican Sisters' first school and the Colégio Sagrado Coraçao de Jesus, founded by the Dominican Sisters.

Photo 2 - Caetanato, the first space where the Dominican School operated

Photo 3 - Colégio Sagrado Coraçao de Jesus Founded by the Dominican Sisters in the 1904s

Porto Nacional's transportation and communication system was linked to the Tocantins River, which was navigated by boats propelled by rowers or vareiros. In 1923, the first steamboat was launched on the Tocantins and, in 1929, the first two motor vehicles arrived in the city: a truck and a car. Photo 4 shows the truck that arrived in the city.

Photo 4 - First truck to arrive in the city in 1929

Source: Author's personal archive.

From the 1930s onwards, the Correio Aèreo Nacional - CAN - developed an air link. This route left Rio de Janeiro and arrived in Belém, landing at the airports set up by Lysias Rodrigues, including Porto Nacional. Photo 5 shows the first plane landing in the city of Porto Nacional.

Photo 5-   First plane to land in the city of Porto Nacional in the 1930s

Source: Author's personal archive.

The city of Porto Nacional is currently considered by the annals of history to be a city of the genesis of social movements, especially during the 19th century. Santos (2004) describes that the 1950s were marked by social movements seeking the dismemberment of the south of Goiás from the north, which is now the state of Tocantins. The struggle for this division of the state was highlighted by the leadership of the people of Porto, who, during the 1960s, started an effervescent movement organized by the Casa do Estudante do Norte Goiano - CENOG, with the aim of overcoming the socio-economic problems that afflicted the population of northern Goiás. The author also adds that, in the mid-1960s and early 1970s, the Belém-Brasilia highway was built, which in a way provided integration, but it wasn't enough to eliminate the region's social problems, of which there were many.

As a result of this popular struggle, the creation of the state of Tocantins was consolidated in 1988. The city of Porto Nacional, along with Araguaina and Gurupi, fought to become the provisional capital, which would house the state powers until the state capital - Palmas - was built. Araguaina and Gurupi were confident about the possibility of being the capital, a fact that was not confirmed when Miracema do Norte was announced as the provisional capital for a year.

Although it is not the capital, Porto Nacional has become one of the state's most important historical references due to the historic settlement of the north of Goiás and, later, the struggle for its dismemberment. Its historic center has been listed by the National Historical and Artistic Heritage Institute (IPHAN) and contains approximately 250 heritage buildings. IPHAN justifies this listing on the basis of studies that prove that the town's origins date back to the Gold Cycle in the 18th century.

New perspectives have emerged as alternatives for the municipality of Porto Nacional: the valorization of its heritage, tourism, commerce; the installation of technical and higher education institutions from the federal and private networks; and agricultural activities. These prospects have become the region's new reality.

2. URBANIZATION, URBANISM AND URBAN PLANNING IN BRAZIL AND TOCANTINS

In order to understand the urban phenomenon and its particularities, a historical conceptual approach to the process of urbanization, urbanism and urban planning itself is essential. It is also important to have an understanding of the role of the municipality in the Federal Constitution and infra-constitutional laws, as well as the regional reality in which it develops. All these elements are considered factors of great relevance to this research, as they are the basis for discussing the processes of occupation, transformation, planning and management of urban space, the focus of the study.

Castells (2009) discusses the historical process of world urbanization, debunking the idea that the genesis of urbanization is a consequence of industrialization. He also states that this embryonic process occurred as a result of the bourgeoisie's independence from feudalism and, as a result of this fragmentation, there was a major transformation in political-administrative and social structures, which could not be followed by all countries in a homogeneous way.

According to Coriolano (2011), in Brazil, the population density of cities from the beginning of the 20th century and the consequent emergence of urban problems were the main motivators for territorial planning. And planning followed international trends, mainly European, of urban intervention, based on hygiene and beautification. In Brazil, at the end of the 19th century and the beginning of the 20th century, urban planning was influenced by English urbanism, especially Howard's garden cities, civic-embellishment urbanism and sanitary urbanism, which aimed to guarantee environmental health. The hygiene movement in the first half of the 19th century gave rise to models of territorial organization/urban design aimed at creating and ordering cities.

In order to discuss the urban planning process in Brazil and Tocantins, it is necessary to understand the theoretical foundations that have guided urban planning since its inception and its influence on the production of cities in the world and in Brazil. According to Coriolano (2011), cities began to see a series of urban problems (lack of sanitation, disease, high density, environmental degradation, etc.), which, combined with pressure from the bourgeoisie, forced the state to become concerned with the planning and management of cities, thus giving rise to the first proposals and attempts at urban intervention in order to solve the problems encountered.

In this sense, the aim of this chapter is to develop a debate on urbanization, urbanism and urban planning, and therefore their genesis in the world and in Brazil. The theoretical and conceptual discussion of the process of global urbanization, with a focus on Brazil, will be discussed.

2.1 URBANIZATION

When it comes to the process of urbanization in a historical approach, Spósito (1988, p.11) considers that urbanization - as a process - and the city - as the concrete form of this process - mark contemporary civilization so profoundly that it is often difficult to think that, at any time in history, cities did not exist, or played an insignificant role.

Spósito (1988, p. 12) also states that "the city of the dead precedes the city of the living". Despite the fact that this was a period marked by people not settling down (nomadism), the dead were regarded with respect by Paleolithic man and had to have a permanent home. The author considers the village to have preceded the city and cannot be considered urban, since the activities related to it were agriculture and livestock, thus differentiating the rural from the urban.

In this context, the city and the countryside differed explicitly in the past. Lewis Munford (1965, p. 13) analyzes that,

Before the city, there was the small settlement, the sanctuary and the village; before the village, the camp, the hideout, the cave, the heap of stones; and before all that, there was a certain disposition towards social life that man shares, of course, with various other animal species.

Spósito (1988) states that there is no precise moment that marks the origin of cities. Probably around 3500 BC, cities appeared in the Mesopotamian region and near the valleys of the Nile, Indus and Yellow rivers. Thus, their origins are directly related to social and political factors.

As a political factor, the author cites the Roman Empire as the best example of the expansion of urbanization in antiquity, due to its unified power. She also mentions that the political power of the Roman Empire allowed the process of urbanization to stop being a spontaneous process, since many cities were founded in the newly conquered areas to allow political hegemony over these areas.

As for social organization, it was marked by: (i) the specialization of work, reflected in the social and territorial division of labour, (ii) the fact that cities were a space of political domination, (iii) the increase in food production and distribution capacity, (iv) writing as a means of maintaining political power, and (v) the internal organization of the urban space, which reflected the social and political structure, in which the center was the place of social institutions, political power and the elites, and around it were the craftsmen and, further away, the agricultural producers (SPÓSITO, 1988).

In the feudal period, there was a decline in cities, in which "land became the only source of subsistence and the condition of wealth" (SPÓSITO, 1988, p. 27-8). In this circumstance, artisanal production, previously produced in the city, returned to the countryside, and social organization turned to land ownership, with an almost exclusively agricultural mode of production.

According to R. Bastide (1978) quoted by Santos (2009), it was only from the 18th century onwards that urbanization developed and the town house became the most important dwelling for the farmer or mill owner. It took a century for urbanization to reach maturity, and even more than a century for it to reach its current state.

Hauser and Schnore (1975) consider that in many Asian, African and Latin American countries, cities are the result of their colonial heritage, as well as embryonic economic development, in which a large part of the population comes from troubled and insecure rural areas. Paradoxically, in the developed countries of the north, especially in Europe, urbanization is old and has been organized slowly at the pace of economic development, which in a way has helped to minimize the atrophies in the social development of these countries.

According to Castells (2009), there are two concepts of urbanization: the first is the spatial concentration of a population, within certain limits and at a certain density; and the second is also called urban culture, which is the diffusion of a system of values, attitudes and behaviours, characteristic of capitalist industrial society. He also establishes theoretical and empirical links between spatial and cultural forms, characterizing urbanization as a process of population concentration on two levels: the proliferation of concentration points and the increase in the size of each of these points. In this sense, Spósito (1988, p. 14) infers that "embedded in the origin of the city there is another differentiation, the social one: it requires a complexity of social organization only possible with the division of labour". Likewise, Hauser and Schnore (1975, p.489) consider that cities are products or consequences of urbanization, "so urbanization is a social process that necessarily precedes and accompanies the formation of cities".

In this sense, Hauser and Schnore (1975, p. 491) study the history of urbanization **in** two phases, calling them **incipient urbanization** and **effective urbanization**. They consider the first to be "the result of a primary urban organization as an additional, more productive system of collective adaptation to the physical and social environment", and the second to be the definitive city, as a result of its capacity to produce, store or use social reserves, with a capacity to rise above its territorial limits, thus resulting in rates and levels of urbanization before and after 1700. Thus, two more categories were created: **classical urbanization**, in which the control of population growth and city dwellers was controlled by social measures of repression and balance; and **industrial urbanization**, which favoured an unprecedented concentration of population, by minimizing the previous category, "the industrial-urban city projects outwards from its native environment into even wider swathes of physical and social environment" (HAUSER; SCHNORE, 1975, p. 491).

Hauser and Schnore (1975, p. 45) also point out that studies on urbanization have a multidisciplinary character, in which "their members have been recruited from the fields of economics, geography, history, political science, social anthropology and sociology".

Manuel Castells (2009, p.16-17) states that the culturalist tendency to analyse urbanization is based on a link between a certain technical type of production and a system of values and a specific form of organization. The link between spatial form and cultural content cannot be considered a defining element of urbanization, although it is a hypothesis. Establishing theoretical and empirical links between spatial and cultural forms.

In this way, Castells (2009) provides a characterization of imperial cities, exemplified by Rome, which represented the political-administrative apparatus through management and domination. And in the Middle Ages, it was based around a pre-existing fortress and the market, especially from new trade routes, representing the emancipation of the bourgeoisie from feudalism.

Otherwise, a new economic and social structure began to be organized, which depended on its financial and manufacturing capacity.

Industrial capitalism weakens the development of cities by organizing itself around specified objectives, resting on the foundations of the weakening of agrarian structures and the emigration of the population to existing centers, the emergence of the domestic economy for manufacturing and, later, for the factory, the market and the industrial environment. Cities act as industrial attractions because they offer labor and a market, while industries bring job opportunities and demand for services. However, "urban disorder" can be seen as a consequence of the rapid pace of urbanization in the world, the concentration of urban growth in underdeveloped regions, the emergence of new urban forms and the relationship between the urban phenomenon and new forms of social articulation stemming from the capitalist mode of production.

Castells himself (2009) also emphasizes that urbanization constitutes specific spatial forms of human societies with: the characterization of the existence and diffusion of a specific cultural system, which is urban culture; and the notion of urban belongs to the ideological dichotomy of traditional society/modern society, in which the city differs from the village in terms of the spatial forms of social organization. The ideological notion of urbanization refers to the process by which the population concentrates on a certain space. The analysis of urbanization is linked to development, referring to a technical and economic level of transformation of social structures. And the problem of development is the transformation of the social structure in relation to investment/consumption. Thus, the notion of development must be articulated with that of dependency, characterized by its asymmetrical social relations. This occurs through the guiding question: what is the process of social production of the spatial forms of a society and, in turn, what are the relations between the constituted space and the structural transformations of a society, within an intersocial set characterized by relations of dependence?

Another important issue to be analyzed with regard to the process of urbanization is Brazilian urbanization. Milton Santos (2009), in his work *Brazilian Urbanization,* discusses the genesis of towns and cities in Brazil in the 16th, 17th and 18th centuries, and their political and economic organization, calling them the social system of the colonies. The elements present in them are: political and administrative organization, rural economic activities and the corresponding social strata, urban economic activities and their actors. He also states that it was only from the 18th century onwards that urbanization developed, and the town house became the most important dwelling for the farmer or mill owner. He points out that the expansion of commercial agriculture and mineral exploration formed the basis of settlement and the creation of wealth, leading to the emergence of cities on the coast and in the interior. So much so that the Brazilian population grew from 9.9 million to 14.3 million in just fifteen years, and it was only after 1940 that the urban population began to be separated from the rural population of the same municipality.

According to Milton Santos (2009), while urbanization grew by less than four points in thirty years (1890-1920), it only took twenty years (1920-1940) for this rate to triple and the urban population to rise from 4.552 million in 1920 to 6.208 million in 1940. We can see that the country was made up of sub-spaces that evolved according to their own logic, largely dictated by their relationship with the outside world. From the second half of the 19th century, through production, the southern region, including Rio de Janeiro and Minas Gerais, became a dynamic hub, which can be explained by the changes that took place in the engineering and social systems. This fact reveals a sub-space that differs greatly from the rest of Brazil, favoring the development of the industrialization process, especially in Sao Paulo.

Milton Santos (2009) also presents data from 1970 onwards, in which the process of urbanization reached a new quantitative and qualitative level, with a reflection of the past in the distribution of urban population agglomerations in each region. The author is critical of the term "medium-sized cities", because what was considered in the 1950s is different from 1970, which is different from today.

In this context, Milton Santos (2009) deals with the **Technical-Scientific Environment**, which is represented by the current phase of construction or reconstruction of space through the content of science, techniques and information, replacing the natural environment. It is understood that the territory is becoming computerized faster than the economy or society, a fact that promotes two distinct situations: the ease of knowing the territory and the excess of information, which confuses the understanding of space.

In this sense, another important point to be discussed is the settlement of the territory of the former north of Goiás, now the state of Tocantins, which, according to Pinto (2012), has undergone several changes throughout its history, as part of a national logic of territorial occupation that relegates the central north of the country to a subordinate role in the division of labor. Lira (2011, p. 122) considers that this territory is characterized as a "food producer, standing out for its vast arable land and livestock", and far from the major urban centers and industrial development. The author divides the occupation of northern Goiás into three periods:

- mining period: discovery of gold (1725) and navigation of the Tocantins;

- republican period: the railroad [Estrada de Ferro Goiàs], the 'March to the West';

- pre- and post-64 period: Brasilia, Belém-Brasilia and the "Legal Amazon".

Aquino (1996, p. 29-30) tries to elaborate on these periods of occupation by identifying other factors that also contributed to the urbanization and occupation of the territory, such as

[...] the auriferous period (18th century); traditional farming (19th-20th centuries); spontaneous and official colonization in pioneer areas (first decades of the 20th century), as well as the crystal mines, which gave rise to some cities in the north (first half of the century): Cristalândia, Pium and Dueré. In addition to these, they also contributed to the birth of some cities, such as: Military Prisons (Araguacema); Villages (Dianòpolis, Pedro Afonso, Itacaja and Tocantinia).

The model of urbanization that emerged was "oriented" occupation along the axis of regional integration, in the Tocantins River valley (Lira, 1995, p. 21), thus the "port" cities emerged or developed, such as Porto Nacional, Pedro Afonso, Miracema do Norte, Tocantinia, among others, which interfaced with Belém, in Parà, and Carolina, in Maranhao. Aquino (1996, p. 52) states that

The north remained isolated until the middle of this century [the 20th century], creating a huge demographic vacuum, with a small number of existing towns, predominantly on the right bank of the Tocantins River, most of which originated during the gold rush. These are known as traditional towns. What's more, the urban life that existed at that time subtly followed the contours of the Tocantins River.

Pinto (2012) refers to those cities that had been developing on the banks of the Tocantins River and which are suffering major impacts from these changes. He mentions the city of Porto Nacional, which, according to him, at the moment still maintains a certain ascendancy in the region, also benefiting from the construction of the bridge over the Tocantins River in 1979, linking the BR-153 highway with the "corridor of misery", which was the right bank of the river.

2.2 URBANISM

The elaboration of this chapter includes a discussion supported by the theorists Lefebvre (2009) and Souza (2005; 2010) about urbanism, a phenomenon that takes place after urbanization, resulting from urban management and planning. These two themes (urbanization and urbanism) are directly linked to social adversities and the supremacy of city management, carried out by public and private service actors.

As previously discussed, Lefebvre (2009) confirms that there were ancient cities: oriental and archaic, both of which were essentially political; and the medieval city, which integrated the commercial, artisanal and banking character with the political.

Like urban planning, urbanism dates back to the Middle Ages, but in an unexplained way, with defined terminology and concepts, as it was much more practical than theoretical. According to Benevolo (1971), it can be said that modern urbanism was born even before the term was used, between 1830 and 1850. Thus, Abiko et al. (1995) state that the term urbanism is relatively recent, and according to G. Bardet quoted by Abiko et al. (1995), the term first appeared in 1910. Abiko et al. (1995) infer that the industrial city of that period was characterized by congestion and insalubrity, with no water supply or sewage system and no garbage collection for the working class population. This led to epidemics that were difficult to control, as well as diseases that affected the population as a whole.

For the same author, this city is built by the private sector, seeking maximum profit and exploitation, without any control. The need arose for public action, ordering and proposing solutions that until now had only been implemented by the private sector, with individual, short-term and small-scale objectives. This was the era of sanitary urbanism, basically concerned with improving health conditions in cities, coordinating private initiative with public and general objectives.

Abiko and others (1995) confirm that health laws have evolved into legislation specifically of an urban nature, defining densities, criteria for the implantation of allotments, distance between buildings, their height gauge, and even the characteristics of each building, i.e. spaces, openings and materials to be used.

Today's urban planning regulations, zoning laws, land use and occupation and building codes all stem from this sanitary concern to create a healthy and suitable environment. At the level of ideas, the first intellectuals to study and propose ways of correcting the ills of the industrial city were polarized in two extremes: either they defended the need to start again from the beginning, contrasting the existing city with new forms of coexistence dictated exclusively by theory, or they tried to solve the singular problems and remedy the inconveniences in isolation, without taking into account their connections and without having a global vision of the new city organism.

Thus, Abiko and others (1995) say that the first case includes the so-called utopians - Owen, Saint-Simon, Fourier, Cabet, Godin - who do not limit themselves to describing their ideal city, but are committed to putting it into practice; In the second case, there are the specialists and officials who introduce new hygiene regulations and new facilities into the city and who, having to find the technical and legal means to carry out these changes, effectively initiate modern urban planning legislation. The utopian urbanists gave rise to an anti-urban stance that opposed industrialization, with the emergence of proposals for garden cities.

In this sense, Lefebvre (2009, p. 139-149) attributes utopian urbanism to a variety of influences "from humanists, real estate developers, the state and technocrats". The author considers that urbanism implies the fetishism of satisfaction, neglects social needs, implies contradictions, "because it is a contemporary political and scientific thought" that needs to break with models.

Abiko et al. (1995) discuss the Charter of Athens, stating that the document was based on the premise that the transformation of social structures and the economic order should correspond to the transformation of the architectural phenomenon. It is inferred that urbanism was one of the keys to a qualitative change in society and human life. These experts particularly emphasize the fact that "building is an elementary human activity, closely linked to the evolution of life. One of the main foundations of the Athens Charter states that the "city is part of an economic, social and political whole, which constitutes the Region. No urban planning problem can be approached without constant reference to the constituent elements of the Region". With regard to town planning, the Athens Charter states that it cannot be subject to the rules of a gratuitous and sterilizing aestheticism. Urban planning must, by its very essence, be functional.

For Abiko and others (1995), according to the Charter of Athens, the city has four fundamental functions for which urbanism must take care: to live, to work, to circulate and to cultivate the body and spirit, with its objectives being land occupation, the organization of circulation and legislation. The Charter of Athens summarizes the content of Rationalist Urbanism, also known as Functionalist Urbanism, which presupposed the obligation of regional and intra-urban planning, the submission of private ownership of urban land to collective interests, the industrialization of components and the standardization of constructions, concentrated building, but adequately related to large areas of vegetation. It also admits the intensive use of modern technology in the organization of cities, functional zoning, the separation of vehicular and pedestrian circulation, the elimination of the corridor street and a geometric aesthetic.

Lefebvre (2009, p. 139) defines urbanism "as the activity that traces the ordering of urban establishments in the territory with stone, cement or metal strokes".

For Goitia (1992), official bodies, planners and urban designers are slow to predict and even slower to realize. While they delimit convenient areas and plan on their basis, preparing solutions for growth, reality, with its violent imperatives, breaks through in the most unforeseen and incongruous places. And when the authorities decide to take them into account, they are faced with an unpleasant and voluminous reality, which changes the data of a problem that was projected to be addressed serenely on the drawing boards. The city is being transformed by growth that is neither orderly by technical means, nor slow and organic by natural means.

The city is notably a space marked by tensions and conflicts that portray social inequalities and urban problems. Erminia Maricato (1997, p. 42) states that

Urban spaces are not just places or stages for industrial production, the exchange of goods, or places where workers live. They are all this and much more; they are products: buildings, viaducts, streets, signs, posts, trees, in short, landscapes that are produced and appropriated under certain social relations. The city is both an object and an active agent of social relations.

Urban planning, therefore, has the task of identifying the real needs of the city in order to come up with feasible solutions. It must put existing social relations at the forefront of its concerns, which is often not the case, i.e. urban planning measures end up straining social relations. Hely Lopes Meirelles (1993, p. 379) summarizes well the primary task of urban planning, which is to solve the problems and conflicts that occur in the city, as he considers that

Urban planning is the set of state measures designed to organize living spaces in order to provide better living conditions for people in the community. Living spaces are understood to be all the areas in which people collectively exercise any of the four social functions: housing, work, circulation and recreation.

In fact, these social functions of the city are diffuse interests, i.e. those of the entire community, whose subjects are not determined. In order to fulfill its social functions, the city must guarantee all citizens, without distinction, the right to an individual and collective guarantee of the environment, housing, urban land, sanitation and infrastructure, transportation and public services, work and leisure, both for present and future generations.

As far as Brazilian urban planning is concerned, the 1988 Constitution's innovation in some aspects, including the inclusion of articles 182 and 183 in the Urban Policy chapter, was theoretically a victory for the active participation of civil organizations and social movements in defense of the right to the city, housing and access to better public services.

Pereira (2002, p. 3), in establishing the relationship between the City Statute and current urban planning practice, gives an idea of what repercussions this law could have on the implications of planning and managing urban space,

[...] in this way, urbanism moves from a technical-projective practice, where urban planners defined the destiny of the city based on a few theoretical elements, to a practice with much more political content, where the central issue is no longer the space of the cities, but the inhabitants of the cities. We've gone from imposing urbanism to proposing urbanism. We have moved from a technocratic urbanism of certainties to a democratic urbanism of possibilities.

Bernardi (2006) states that the City Statute is therefore the result of a long historical-political process, which only developed as a result of popular movements. It is a law that takes into account the reality experienced by the population and not an idea to be pursued. On the other hand, the City Statute should only be seen as an instrument and not as a solution, which should be sought by the social agents involved. It is up to the municipalities to regulate the Statute's instruments in order to ensure that cities are built based on values of social justice. It is necessary to seek harmony between general interests and private interests within the urban space and not to favor specific interests. The action of the population is essential in this process.

Some of the instruments contained in the City Statute are specific to the planning of urban land, such as zoning, which guides the preparation of plans, programs and/or projects; financial tax institutes, which promote the contribution of improvements, tax and financial incentives; and legal and political institutes, such as: expropriation, tombamento, establishment of conservation units, establishment of spatial zones of Social Interest - ZEIS, usucapiao, outorga onerosa, direito a preempçao, operaçoes urbanas consorciadas and others, and the study of Neighborhood Impact. It is extremely important for city dwellers to know that these instruments exist and that, if practiced, they can profoundly change their living conditions.

Bearing in mind that these instruments are the subject of this research, it is worth highlighting the importance of the Municipal Master Plan as an instrument of urban policy.

According to Saule Jûnior (1997, p. 22), this right to a sustainable city means

[...] it includes the inherent rights of people living in cities to have decent living conditions, to exercise their citizenship peacefully, to broaden their fundamental rights (individual, economic, social, political and environmental), to participate in city management, and to live in an ecologically balanced and sustainable environment.

What can be said, therefore, is that cities must also fulfill certain social functions for the well-being of the community. And these social functions of the city are linked to the environment, which ensures the quality of life of people living in urban settlements.

In this context, it is necessary to focus on the socio-spatial object of the research, which is the city, so the discussion is redirected to the Municipal Master Plan. In this case, the Porto Nacional Sustainable Development Master Plan (PDDSPN) is no different from any other. Coriolano (2011), in his State Report on the Evaluation of Participatory Master Plans in the State of Tocantins, analyzes the municipalities of Araguaina, Gurupi, Palmas and Porto Nacional, at which point he discusses this document. It also states that the general duties of the Master Plans are to create and maintain areas of special environmental interest and to restrict the use of areas at geological risk. One of the specific guidelines is the protection, preservation and restoration of the natural and built environment. However, it lays down generic actions in just one article, such as the protection of water resources, the enhancement of APP's and UC's, the recovery of springs in the urban area, etc. It also suggests that the municipality should formulate the Municipal Environmental Policy and the Municipal Environmental Code. The Plan establishes the Macrozoning of the Municipality, creating the Environmental Protection Macrozone (MA) consisting of the Palmas Lake Environmental Protection Area.

In this context, Santos Júnior (2011, p. 42) states that,

[...] despite the large number of Master Plans that incorporate the environmental issue among the fundamental objectives and principles of urban development policy, especially through the concept of sustainability and environmental quality, few incorporate mechanisms and instruments capable of making environmental policy effective.

Article 26 of Porto Nacional's Sustainable Development Master Plan law discusses the concept of macro-zoning as the division of the municipality's territory into integrated areas. The aim is to promote its planning, as well as the planning and proper implementation of the strategic lines and action programs defined therein.

Article 27 of the aforementioned document, as presented in the introduction to this work, establishes Porto Nacional in the following macro-zones: (i) Urban Macrozones (MU); (ii) Environmental Protection Macrozone (MA); and (iii) Rural Macrozone (MR). The area spatially delimited for this study will be covered by the first two macro-zones, i.e. the Urban Macrozone (MU) and the Environmental Protection Macrozone (MA).

Article 67 of the Porto Nacional Sustainable Development Master Plan law, in item VII, delimits the landscape-environmental area, corresponding to: the area destined for the preservation of the natural landscape on the edge of the lake and its slopes for activities linked to tourism and leisure; permanent preservation areas, destined for the preservation of natural areas, comprising the banks of the Sao Joao stream.

Given this context, we return to the statement that this research is aimed at studying the theoretical and conceptual foundations, as well as the legal frameworks of Brazilian urban environmental policies, in order to gather information that will answer its research question. In this way, the research will go on to assess the degree of effectiveness of the guidelines of Porto Nacional's Sustainable Development Master Plan as an environmental management instrument, applied in plans, programs and projects in a way that corroborates the well-being of the city's inhabitants.

2.3 URBAN PLANNING IN BRAZIL AND TOCANTINS

In order to approach urban planning, it is necessary to first conceptualize planning, preceded by a historical analysis. Thus, Bernardi (2006) indicates that planning means directing, anticipating the future in the present moment. Aguiar (1996, p. 35) argues that "planning means establishing objectives, indicating guidelines, studying programs, choosing the most appropriate means for achieving them and outlining the government's actions, taking into account the possible alternatives".

Therefore, planning is not a natural and irrational action. It is, therefore, something that is methodically systematized in a thoughtful and deliberate way, with rules and standards to be followed, in all areas in which it is applied, and especially in urban planning. The actions of planning and managing are apparently interlinked, so Souza (2010) warns us that in Brazil, since 1985, the use of the expressions urban management, environmental management, territorial management, educational management, science and technology management, among others, has intensified, but with a neoliberal ideological slant.

Souza (2010, p. 46) states that planning and management have different time frames and refer to different types of activities, while

[...] planning always refers to the future [...] management refers to the present. [...] it is the preparation for future management, seeking to avoid or minimize problems and expand margins for manoeuvre; and management is the implementation, at least in part, of the conditions that past planning helped to build.

In this circumstance, it can be inferred that planning precedes management, but they differ in terms of the execution of activities. With regard to urban planning, Souza (2010) states that it requires various professionals, such as architects, social scientists from different backgrounds, especially geographers, and other professionals, such as specialists in urban law, demystifying the idea that only architects are the protagonists of this activity.

Following a historical outline of the origins of urban planning, Bernardi (2006) states that Brazilian urban planning dates back to the colonial period, as some Brazilian cities, such as Sao Vicente (1532) and Salvador (1539), were planned and dated back to their foundations. Tomé de Souza had a team of architects, surveyors, bricklayers and carpenters, among other professionals, to lay the foundations of the country's first capital. What is clear is that this planning was less explicit in its intentions, as David Clark (1991) argues that urban planning was first recognized at the end of the 19th century and the beginning of the 20th century in the United Kingdom, Europe and North America, when the problems of industrial cities were evident in various ways, including through poverty and disease.

Santos Júnior (2011, p. 48) points out that "Brazilian cities did not have their longest period of growth guided by an urban planning process", which the author justifies by the lack of housing and infrastructure present in Brazilian cities.

Souza (2005, p. 274) also states that "the banner of urban reform is not a recent one in Brazilian history", as it received its visibility in 1963, when the Peasants' League movement was at its peak and the lack of housing was a notorious urban problem, due to the lack of urban planning in Brazil.

And for Souza (2005, p. 274), it is necessary to go back in time to better understand the urban planning that takes place through the clamor for urban reform. The author goes back to the 1970s and states that the mid-1970s were marked by the "political opening" implemented by General-

President Ernesto Geisel. A period marked by

[...] the channeling of the clamor for more social justice in the city through residents' associations by residents of favelas, clandestine subdivisions and other poor residential spaces, and even by middle-class residents mobilized for a better quality of life and against the aggressions of real estate capital, prepared the ground, so to speak, for urban reform.

Souza (2005) goes on to say that in the 1980s, these municipal and state federations of associations began to deal with problems that had previously been specific, but were now more comprehensive at city and even country level, acquiring a properly political content. In this way, municipal and state federations of associations began to interact with other civil society organizations, some unions, NGOs, universities and others, contributing to the implementation of the legal instruments of urban reform through the new Brazilian Magna Carta in 1988.

Articles 182 and 183 of the Federal Constitution regulate urban planning and control:

Art. 182. The aim of urban development policy, implemented by the municipal government in accordance with the general guidelines laid down by law, shall be to organize the full development of the social functions of the city and guarantee the well-being of its inhabitants.

Art. 183. Anyone who owns, as their own, an urban area of up to two hundred and fifty square meters, for five years, uninterruptedly and without opposition, using it for their home or that of their family, will acquire ownership of it, provided that they do not own another urban or rural property (BRASIL, 2002).

The approval of this law is considered by Santos Júnior (2011, p.48) to be the promotion of an urban development and expansion policy, with the aim of organizing the social functions of the city and guaranteeing the well-being of city dwellers, stating that

[...] the legal framework for urban policy at the national level constitutes a great opportunity for urban planning to be strengthened and structured in municipalities, so as to contribute to better use of public resources, to maximize their effects on the city and to reduce social and urban infrastructure deficits.

Redirecting this discussion to urban planning in Tocantins, it is important to point out that the State of Tocantins was created from the division of the State of Goiás in 1988, through Article 13 of the Transitional Provisions of the new Federal Constitution. After an intense popular struggle in the north of Goiás, the new state was institutionalized and a new city was proposed to house the capital. However, it should be noted that until this event was consolidated, other important attempts throughout history were organized, and although they did not achieve the success desired by their leaders, they became fundamental for understanding some aspects present in the construction of the state and the layout of its capital, considered a planned city.

As far as the settlement of the region is concerned, throughout the history of northern Goiás, now Tocantins, the BR-153 highway has played an important role. This is because several towns have sprung up along this highway, many where the construction sites of the construction companies were located, others around commercial stops, and even towns that moved from slightly further away to its banks. Paraiso do Tocantins is an example of these towns, which Lira (2010, p. 224) calls "camp towns". While Oliveira (2009, p. 69) considers that "the territorial organization of the state of Tocantins", formerly known as Norte Goiano, "had and has a direct influence on the expansion policies of the country's agricultural frontier". The author refers to those cities that had been developing on the banks of the Tocantins River, and which suffered major impacts from these changes. An example of this is the city of Porto Nacional, which at the moment still maintains a certain ascendancy in the region and also benefited from the construction of the bridge over the Tocantins River in 1979, linking the BR-153 highway with the "corridor of misery", which was the right bank of the river.

According to Velasques (2009), with the end of the military regime in 1985, the National Constituent Assembly was formed to draw up a new constitution for Brazil. In this favorable context, there was a complex political articulation in favor of the division of the state, not only in Brasilia, among the deputies and senators, but also in the executive power of the state of Goiás, with the support of the then governor Henrique Santillo for the separatist proposal, as rapporteur of the Subcommittee of the States in the National Constituent Assembly. After the Legislative Assembly approved the proposal to build a new city, the then governor of the state (Siqueira Campos) asked the IBGE to carry out studies to choose the area for the new state capital. Subsequently, he hired the Grupo Quatro Architecture, Urbanism and Planning Office, based in Goiânia, to finalize the studies, define the best location and draw up the project for the new city (VESLASQUES 2009, p.11).

And on May 20, 1989, Governor Siqueira Campos laid the foundation stone of the future capital in the Canela District, which belonged to the newly-emancipated municipality of Taquaruçu do Porto. Lira (2010, p. 230) states that

Palmas comes as the 'new' in this Latin context, but it is in Brazil that it emerges as an old/new model of city, old in the sense of administrative planning, new in the sense of being a capital built by 'private' capital, and also because it is the newest frontier of urban capital in the Brazilian space.

The studies carried out for the choice of the capital (1989) defined a large quadrilateral in the central region of the state, identifying possible areas for the installation of the city, including the Canela region, where the capital was to be installed (PINTO, 2012, p. 87). In this regard, Bessa (2011, p. 8) states that

Palmas [...] was designed to be the capital, located in the geographical center of the new state, with the aim of moving traffic away from the Belém-Brasilia axis, towards the less economically dense areas of the former north of Goiás, and signaling, on the one hand, a movement to overcome the unequal spatial configuration induced by the BR-153, and, on the other hand, to a movement to affirm internal political forces in opposition to external forces, seeking autonomy in managing the processes of consolidating its territorial structure.

Pinto (2012, p. 86) argues that the city is part of an archaic model of Brazilian urban development in terms of the process of incorporation, appropriation and transformation of space. Lira (2010, p. 284) agrees with the author and states that

In this sense, the occurrence of urban planning in Brazil and in Tocantins is directly associated with numerous factors, including economic development policies, which are reflected in major social inequalities. In the state of Tocantins, as in other Brazilian states, there is no urban planning based on social practices, which promotes the economic and social development of cities in a contemporary way. On the contrary, they have been locked into their own development plans.

3. THE POLITICAL AND LEGAL INSTRUMENTS OF URBAN PLANNING IN BRAZIL

This chapter will discuss the legalization of urban planning in Brazil, including social movements, the 1988 Constitution, the City Statute and Master Plans, using the theoretical contributions of Souza (2010), Ribeiro and Cardoso (2003), Gasparini (2002) and others.

The Constitution of the Federative Republic of Brazil, approved in 1988, also known as the "Citizen Constitution", is considered by many researchers to be the first legal framework for urban policy in Brazil. This important legal instrument proposed the decentralization of the Federal Government's administrative power, passing on to the municipalities the responsibilities that had previously belonged to the Union.

In the context of the drafting of the 1988 Federal Constitution, the 1980s saw a major mobilization of popular movements in Brazil to tackle the social inequalities and urban problems that had been on the agenda since the previous decade. Among those who organized themselves mainly in the context of the drafting of the Constitution and in search of the establishment of social rights, was the National Movement for Urban Reform - MNRU, made up of non-governmental organizations, class entities, housing movements and trade unions. This movement presented Popular Amendment No. 63/1987, known as the "Popular Amendment for Urban Reform", signed with 131,000 signatures (SOUZA, 2010, p. 159).

Although the amendment proposed by the MNRU was not fully incorporated into the constitutional text, it was nonetheless relevant to the establishment of an urban policy in the country with a proposal for urban reform and fairer use of the territory. In this way, it has come to influence an important urban planning instrument, as discussed in the next chapter.

3.1 SOCIAL MOVEMENTS AND THE URBAN REFORM CHAPTER OF THE 1988 CONSTITUTION

In order to understand the legalization of urban planning in Brazil, it is necessary to understand the social movements that contributed decisively to the creation of the Urban Reform chapter in the 1988 Constitution. This chapter of the Constitution has the function of reducing the levels of social injustice in the urban environment in order to promote greater democratization in the planning and management of cities.

In this sense, Souza (2010) argues that the appropriation of the idea of urban reform through urban planning and management by critical thinking did not originate in the First World, but in Brazil. The idea of urban reform from a progressive perspective seems to date back to the 1960s, although this expression is much older than that. The same author criticizes the de-characterization of the movement, which used to be left-wing, into an adherence to state interventions with an anti-grassroots content.

According to Souza (2010, p. 156), other Latin American countries, such as Colombia, for example, took part in these discussions, but "[...] it should not be thought that discussions on urban reform were the exclusive privilege of Brazilians. [...] the Brazilian debate did not take place in isolation and oblivious to what was happening in other countries".

the author takes a look back at the discussion on urban reform in Brazil initiated by Joao Goulart (1961-1964), who had drawn up an Urban Reform Project, discussed in 1963 at an event in the city of Petrópolis (RJ). However, the effervescent rural agrarian reform movement, the Peasants' League, overshadowed the project's visibility. During the 1960s and 1980s, the idea of urban reform was born. souza (2010, p. 157) states that

This period, from the 1960s to the early or mid-1980s, can be called the "pre-history" of urban reform, because although the core of the idea was already present, with its critical content of seeking greater social justice in urban space, a broader idea, which went well beyond the housing issue and included reflections on instruments, would only be built two decades after the Petrópolis meeting.

Following Souza's approach (2010, p. 158), in the mid-1980s, a period in which the country was awaiting the drafting of a new constitution, urban social needs were growing, including the issue of housing. As a result, the movement's banner diversified and the National Movement for Urban Reform - MNUR - came into being, maturing the progressive conception of urban reform in the country, characterized by presenting

[...] an articulated set of public policies, of a redistributive and universalist nature, aimed at the following primary objective: reducing the levels of social injustice in the urban environment and promoting greater democratization of the planning and management of cities (SOUZA, 2010, p. 158).

It differs in this way from simple urban planning interventions and is equivalent to the "urban version of agrarian reform" (SOUZA, 2010, p. 158), albeit with its limitations inherited from the capitalist social model, since it is a debate about specific programs for generating employment and income, allotments (land use and occupation), cooperatives and self-managed forms of worker organization.

According to Souza (2010), the MNRU solidified and became evident in the second half of the 1980s, with the effective participation of neighborhood activist organizations, professional bodies such as the IAB and AGB, as well as academics. This organization sought an opportunity signaled by the National Congress for civil society to participate in the drafting of popular amendments, with the participation of at least thirty thousand voters and three entities. However, the MNRU succeeded in drafting a popular amendment signed by 130,000 voters.

The National Congress behaved in the face of this exhaustive effort by the MNRU as if it was obliged to receive the proposal, but would not take on the responsibility of incorporating it in its entirety, which led to the document being minimized, excluding proposals relating to public transport, public services and public land from the application of the usucapiao instrument. Only two articles remained - 182 and 183 - with some modifications and quite diluted, which were published in Chapter III, entitled Urban Reform, in the new 1988 Constitution, as transcribed by Souza (2010, p. 159):

Art. 182. The aim of urban development policy, implemented by the municipal government in accordance with the general guidelines laid down by law, shall be to organize the full development of the social functions of the city and guarantee the well-being of its inhabitants.

§ Paragraph 1 - The master plan, approved by the City Council and mandatory for cities with more than twenty thousand inhabitants, is the basic instrument of urban development and expansion policy.

§ Paragraph 2 - Urban property fulfills its social function when it meets the fundamental requirements of city planning expressed in the master plan.

§ Paragraph 3 - Expropriations of urban properties shall be made with prior and fair compensation in cash.

§ Paragraph 4 - The municipal government may, by means of a specific law for the area included in the master plan, demand, under the terms of federal law, that the owner of unbuilt, underutilized or unused urban land promote its adequate use, under penalty of successive fines:

I   - compulsory parceling or building;

II   - progressive urban property tax;

III - expropriation with payment through public debt securities issued in advance and approved by the Federal Senate, with a redemption period of up to ten years, in annual, equal and successive installments, ensuring the real value of the compensation and legal interest.

Art. 183. Anyone who has owned an urban area of up to two hundred and fifty square meters for five years, uninterruptedly and without opposition, using it for their home or that of their family, will acquire ownership, provided they do not own another urban or rural property.

§ Paragraph 1 - The title of ownership and the concession of use shall be granted to a man or a woman, or both, regardless of marital status.

§ Paragraph 2 - This right shall not be recognized to the same possessor more than once.

§ Paragraph 3 - Public property shall not be acquired by usucapiao (FEDERAL CONSTITUTION, 1988).

Oliveira (2001, p. 8) states that

The inclusion of articles 182 and 183 in the Urban Policy chapter was a victory for the active participation of civil organizations and social movements in defense of the right to the city, to housing, to access to better public services and, consequently, to opportunities for a dignified urban life for all.

For the purposes of this research, this approach is fundamental, since the creation of the City Statute law and, consequently, the drafting of master plans in recent times is initially the result of these achievements, albeit with caveats, present in the 1988 Constitution.

3.2   THE CITY STATUTE AND THE URBAN ENVIRONMENTAL ISSUE

The inclusion of the fundamentals of urban reform in articles 182 and 183 of the 1988 Federal Constitution led to the creation of the City Statute (law 10.257/2001). The City Statute has transformed the concept of urban property, since it proposes that real estate should no longer be a way of accumulating wealth, but should have the social function of housing.

Looking at the historical preamble, Ribeiro (2003) states that the City Statute had its first rays in 1976, more than a decade before the 1988 Constitution, when the Brazilian press publicized the existence of a preliminary development project drawn up by the National Council for Urban Policy - CNPU, an organ of the Ministry of the Interior, supported by progressive consultants and technicians. The media at the time raised an intriguing and paradoxical alarm, claiming that the military government intended to socialize urban land. This led to the draft being aborted.

Ribeiro (2003) explains that, as a resistance to this situation, the National Conference of Bishops of Brazil (CNBB) launched the "urban soil" theme of its Fraternity Campaign in 1982, advocating public control over the real estate market as a goal for tackling inequalities in urban living conditions. This initiative sparked a revival of social movements and progressive leaders around a debate on the urban question and its relationship to social justice. It also seeks to discuss a national urban development policy capable of resolving the demands of cities in terms of public services and public tolerance of real estate speculation.

Ribeiro (2003) also reports that, in 1981, the debates of the electoral campaign for state governments, as the first direct elections after the coup d'état in 1964, fermented this discussion. One example is Rio de Janeiro, where Leonel Brizola won the election. His campaign was centered on the proposal "Cada Familia Um Lote" (Every Family a Lot). This proposal presented an urban reform program with which his government intended to settle around 1 million families in the empty lots of the metropolitan region, leading the military government to rescue the old draft and transform it into an urban development bill, as a strategy to disperse the popular layers around the opposition. As a result, the military government's "new" project failed to mobilize parliamentarians, even in the face of intense demonstrations by social leaders, professional institutions, progressive technicians, NGOs and others. Ribeiro (2003, p. 14) states that

All these facts lead us to see that the guidelines, objectives and instruments for regulating land use contained in the City Statute express a solid social and political consensus developed in Brazilian society in this long historical process that began in the 1960s around the urban question and its relevance and centrality in the construction of a project for an equal and just society.

According to Ribeiro (2003), the institutional and political impact of the approval of the City Statute must be assessed taking into account the regulation of the social function of property, which must be translated into master plans.

Santos Júnior and Montandon (2011) argue that this document also brought new directions for urban development by affirming guidelines, principles and instruments aimed at promoting the right to the city and democratic management.

The City Statute is a document made up of 58 (fifty-eight) articles, divided into five chapters: General Guidelines, Urban Policy Instruments, Master Plan, Democratic City Management and General Provisions. In this way, it gives us the responsibility to re-read the provisions in the light of issues relating to the environment, presenting the guidelines and instruments of urban policy that also seek to ensure the protection of the environment. And this is the focus of the research in question, evaluating the urban environmental issue from the point of view of the guidelines regulated by the City Statute, in this case, in particular, the Master Plan.

However, as Milaré points out, cited by Torres (2006), the law does not formally present clear concepts or requirements, much less impose specific obligations for environmental protection, unlike what it did with property and the use of urban land.

Moving on to the discussion of the second preamble presented in this chapter, which sets out to debate urban-environmental issues, it is necessary to delimit the object of analysis. It is therefore important to understand that, in our view, there is no dichotomy between the environment and the urban environment; on the contrary, there is a profound interrelationship between the two that deserves to be addressed.

The rules contained in the City Statute, although linked more directly to urban and not environmental policy, have clear repercussions for the protection not only of the built environment, but also of the natural environment, as expressed in the sole paragraph of Article 1, which states:

Art. 1 [...] Sole Paragraph. For all intents and purposes, this Law, known as the City Statute, establishes rules of public order and social interest that regulate the use of urban property in favor of the collective good, the safety and well-being of citizens, as **well as environmental balance** (BRASIL, 2002) (emphasis added).

Nevertheless, with regard to the lack of immediate treatment in the protection of the environment, Oliveira and Pires Carvalho (2002, p. 57) state that the guidelines of the City Statute clearly set out its precipitous purpose of "organizing national urban development, through the regulation of articles 182 and 183 of the Federal Constitution, with general rules that take into account the collective good, safety, well-being of citizens and environmental balance".

However, the need to pay attention to the urban environment cannot be denied, as Rezende (2003) reminds us. For this author, the visibility of the urban environmental issue is due to the impacts of urbanization, so that

[...] the attempt to use the expression urban environment would, on the other hand, try to unite physical, natural and built aspects of urban space with aspects of urban quality of life, understood as the foundation and synthesis between individual well-being, environmental balance and economic development (REZENDE, 2003, p. 141).

The 1988 Constitution contains a chapter on the environment, guaranteed in Article 225, in which

Everyone has the right to an ecologically balanced environment, which is the property of the people and essential to a healthy quality of life, imposing on the public authorities and the community the duty to defend and preserve it for present and future generations (FEDERAL CONSTITUTION, 1988).

As a result, Rezende (2003) infers that urban environmental issues or the environmental protection of urban areas are not clearly established in the Federal Constitution. The author also points out that, although environmental protection is one of the principles to be observed in economic activities, there is no interconnection between the provisions on the environment and urban policy, especially when it comes to protecting the built environment. There is a tendency to protect it only when it can be considered part of the historical, cultural, artistic or environmental heritage, and the latter is more linked to landscape issues than to built space.

Article 30 of the Federal Constitution gives municipalities the responsibility to "legislate on matters of local interest". In this case, the interests of the municipality in the social and environmental spheres may be feasible, which is why an investigation in this area is necessary in order to assess the extent to which municipal managers have been concerned with these socio-environmental aspects.

However, the approval of the City Statute, Law no. 10.257, of July 10, 2001, after 12 years of the New Federal Constitution, has continued the process of implementing a Brazilian urban policy, including environmental concerns, as stated in Article 2 and item I:

The objective of urban policy is to organize the full development of the social functions of the city and urban property, by means of the following general guidelines:

I-guarantee of the right to sustainable cities, understood as the right to urban land, housing, environmental sanitation, urban infrastructure, transportation and public services, work and leisure, for present and future generations (BRASIL, 2002).

In this sense, Torres (2006) infers that the real interface of the law in the environmental sphere is through planning and control of the use and occupation of urban land. Praised by urban reform movements based on socialist ideas, the City Statute is a faithful product of the current Citizen's Constitution. The relativization of property rights is present throughout the text, in a way that has not been seen in Brazilian law for a long time.

Rezende (2003), in his discussion of the issue in question, states that the City Statute places the Master Plan alongside environmental zoning, raising the level of importance given to environmental issues. He also points out that the Master Plan instrument and land-use planning actions do not automatically include environmental concerns, which are listed in the law as the establishment of conservation units and the prior environmental impact study (EIA), but without much ado. The prior neighborhood impact study (EIV) will establish a more solid link with the built space and the environment.

It is worth remembering that the EIV is one of the research instruments in this study and its proposal is to

[...] evaluate the effects on the population living in a given area as a condition for approving works or the operation of activities. The assessment must be made available to the population and take into account the consequences on population density, urban equipment, traffic generation and demand for public transport, ventilation, lighting, urban landscape and natural and cultural heritage (REZENDE, 2003, p. 149).

The **Neighborhood Impact Study (EIV)** is an instrument found in articles 36 to 38 of the City Statute, which must be prepared prior to the issuance of licenses or authorizations for the construction, expansion or operation of developments that have an actual or potential urban impact. According to Torres (2006), the municipal law that establishes it will require an analysis of the probable effects of implementing certain undertakings or activities, in accordance with the local interest. Both urban and environmental aspects will be taken into account, such as ventilation, lighting, urban landscape and natural and cultural heritage. Article 38 states that the preparation of the EIV does not replace the preparation of the **Environmental Impact Study (EIA)**, required under the terms of environmental legislation.

For Torres (2006), this provision expressly separates what is of urban interest (EIV) from what is environmental (EIA). The latter will be required in accordance with the rules of environmental law, as set out in the City Statute itself. The EIA is required by various environmental laws, but mainly by CONAMA Resolution 01/86. The objectives of each study must be made clear, so that one does not interfere with or replace the other, as each has its own purpose. While the EIV should analyze everyday aspects of the built environment, as already mentioned, the EIA will deal with issues more pertinent to the natural environment, soil morphology, quality of environmental resources, underground springs, fauna, flora, etc. (TORRES 2006, p. 207).

For Rezende (2003), this obligation fills an important gap in legislation and its success lies with the municipalities. The author states that the City Statute establishes a bridge between the environmental/urban fields, benefiting the Master Plans with this synthesis. According to Torres (2006), this law has filled a gap in urban planning, providing the public authorities with a regulatory framework to guarantee the right to sustainable cities.

In fact, based on this premise, Carvalho Filho (2005) states that various environmental guidelines, whether explicit or not, are listed in Article 2 of the City Statute. The sum of these guidelines constitutes the purpose of the execution of Urban Policy by the Public Authorities, with the aim of ensuring the well-being of communities in general.

In addition to the aforementioned guideline set out in item I of art. 2 of the City Statute, others are also consistent with environmental concerns: planning the development of cities, the spatial distribution of the population and the economic activities of the municipality and the territory under its area of influence, in order to avoid and correct the distortions of urban growth and its negative effects on the environment - IV; land use planning and control, in order to avoid deterioration of urbanized areas and environmental pollution and degradation - VI; f) and g) adoption of patterns of production and consumption of goods and services and of urban expansion compatible with the limits of environmental, social and economic sustainability of the Municipality and the territory under its area of influence - VIII; protection, preservation and restoration of the natural and built environment, cultural, historical, artistic, landscape and archaeological heritage - XII; hearing of the municipal government and the interested population in the processes of implementing undertakings or activities with potentially negative effects on the natural or built environment, the comfort or safety of the population - XIII; and land regularization and urbanization of areas occupied by low-income populations through the establishment of special urbanization, land use and occupation and building standards, taking into account the socio-economic situation of the population and environmental standards - XIV (TORRES 2006, p. 202). 202).

In order to comply with these general guidelines, the City Statute listed, in art. 4, a series of instruments of various kinds to be used by municipalities to deal with directly urban and indirectly environmental issues. These instruments are financial taxes, about which Torres (2006, p. 203) mentions that the City Statute did not innovate in this area, because it only mentioned IPTU, improvement contributions, tax and financial incentives and benefits, because

[...] in the same way that the previous instruments force the parceling, building or use of unbuilt, underutilized or unused urban land, the Government can act in the opposite way: by encouraging and stimulating entrepreneurs to build on empty land within the city, rather than on the green areas on the outskirts.

This research aims to investigate the financial or tax instruments used to induce protection of the urban environment in the municipality of Porto Nacional.

Still in compliance with the general guidelines, the City Statute Law, commented on by Torres (2006), emphasizes that the Legal and Political Institutes are broken down into: (i) **tombamento**, in which the government must protect assets of historical, artistic, landscape, tourist, cultural or scientific value, which are an integral part of the built environment; (ii) the **establishment of conservation units**, in which the government can create areas for the preservation of the environment, divided into two distinct groups: Integral Protection Units, which aim to preserve nature, allowing only indirect use of its natural resources, and Sustainable Use Units, to make nature conservation compatible with the sustainable use of part of its natural resources (TORRES, 2006, p. 204).

Disciplined in articles 25 to 27 of the City Statute and as Legal and Political Institutes, Torres (2006) comments on the **(iii) right of pre-emption**, which gives the Public Power the right to acquire the urban property necessary for the purposes set out in art. 26, among which the following are worth mentioning: the creation of public leisure spaces and green areas, the creation of conservation units or the protection of other areas of environmental interest or the protection of other areas of environmental interest. These include: the creation of public leisure spaces and green areas, the creation of conservation units or the protection of other areas of environmental interest or the protection of areas of historical, cultural or landscape interest; and also the **(iv) onerous grant**, which is the possibility of building above the basic utilization coefficient, in return for payment to be made by the beneficiary. Up to the basic coefficient, the full right of ownership is being exercised, which cannot be opposed by the public authorities (except through administrative limitations, easements and expropriation). Above this limit, it is a matter of the right to build, in which the beneficiaries will have to repay the community for the extra added infrastructure and urban services generated, as well as the burden on the environment (population densification, greater generation of pollution, etc.).

Continuing with the discussion of the Legal and Political Institutes of the City Statute and with the debate presented by Torres (2006), **(v) the consortium urban operations**, provided for in arts. 32 to 34, deal with this legal-political instrument, which aims to achieve structural urban transformations, social improvements and environmental enhancement in a given city area, by means of structural urban transformations, social improvements and environmental enhancement. 32 to 34, deal with this legal-political instrument, which aims to achieve, in a given area of the city, structural urban transformations, social improvements and environmental enhancement, through a set of interventions and measures coordinated by the municipal government, with the participation of the residents and owners of the area, permanent users and private investors, local or otherwise. Here, the public authorities will be able to "negotiate" with the other consortium members, among other measures, changes to the indices and characteristics of subdivision, use and occupation of land and subsoil, as well as changes to building regulations, taking into account the environmental impact of these changes.

Environmental zoning is dealt with in the Law as an environmental planning instrument for municipalities, because in addition to macro-zoning, which separates the rural, urban and urban expansion zones of the municipality, and urban zoning, which will regulate the use and occupation of the city's land, the City Statute introduced environmental zoning, which had already been provided for in Law 6.938/81 at federal level, at municipal level. It is an instrument that will enable municipalities to regulate the "occupation and destination of geographical areas so that they meet their geo-economic and ecological vocation" (MILARÉ cited by TORRES, 2006, p. 206). In this context, Torres (2006) also states that the environmental macro-zoning of municipalities should include not only the areas of environmental interest established by the local government, but also state, metropolitan and federal conservation units, given the ubiquitous nature of the environment.

After discussing the institutes and instruments presented in legal terms and aimed at municipal planning, it is pertinent to present the Municipal Master Plan as a means of making them effective. In this sense, Torres (2006) comments that, with the advent of the Federal Constitution of 1988, the Master Plan gained the status of a basic instrument of municipal development and urban expansion policy (§1 of art. 182). Such is its importance that it is now required for all municipalities with a population of over twenty thousand inhabitants, among other requirements. This is because

The Master Plan is based on a reading of the real city, involving themes and issues relating to urban, social, economic and environmental aspects. [...] Its objective is to be an instrument for defining a strategy for immediate intervention, establishing a few clear principles of action for all the agents involved in building the city (BRASIL 2002, p. 40).

In this way, there is no way that the Master Plan can fail to address the environmental issue, to the point that we find the name Sustainable Development Master Plan in the elaboration for the municipality. Franco (1999, p. 22) points out that

[...] these plans must be coherent and synergistic with environmental management plans, since it is impossible to consider the prospects and proposals for an urban area without taking into account its environmental variables. In particular, planning that takes the environment into account should detect points of vulnerability and areas of environmental risk for the settlement of the population and developments, areas targeted for activities which, in turn, can determine different degrees of densification, discontinuities in the urban fabric, axes of expansion and restrictions due to environmental factors, such as watercourses or prevailing wind directions, among many others.

Torres (2006) argues that any issue that interferes with the policy of urban and environmental development and expansion should be covered by the Master Plan. Thus, it is expected to contain at least guidelines and objectives on environmental zoning, especially indicating the areas to be preserved in the face of urban expansion and the areas to be recovered, conservation units and other areas of environmental interest. It should also contain a list of environmental policy instruments, in particular environmental licensing, the municipality's environmental information system and the municipal environmental quality and targets program.

In this way, municipalities have the opportunity to plan their urban development policy together with society, based on the principle of equality and urban sustainability.

According to Pietro (2006), the City Statute provides for various legal, political, tax, financial and study and planning instruments for the organization of urban space. Among the instruments regulated by the City Statute, they can be divided into three categories: plans, institutes and studies, as shown in Chart 1.

Table 1 - Urban Planning and Management Instruments of the City Statute

<table>
<tr><td colspan="2">Framework of the urban planning and management instruments of the City Statute - Law n. 10.257/2001 and subject of this investigation.</td></tr>
<tr><td>1) OF MUNICIPAL PLANNING:
a) master plan;
b) environmental zoning;
c) sectoral plans, programs and projects.</td><td>111) TAX INSTITUTES AND FINANCIAL:
a) property tax - IPTU;
b) improvement contribution;
c) tax and financial incentives and benefits.</td></tr>
<tr><td>11) THE STUDIES
a) prior environmental impact study (EIA); b) prior neighborhood impact study (EIV).</td><td>IV) LEGAL AND POLITICAL INSTITUTES:
a) expropriation;
b) listed buildings or urban furniture;
c) establishment of conservation units;
d) creation of special areas of social interest;
e) concession of real right of use;
f) special adverse possession of urban property;
g) right of pre-emption;
h) onerous granting of the right to build and change of use;
i) consortium urban operations.</td></tr>
</table>

Source: Organized by the author.

The federal City Statute law, as we have seen, in its legal text, has brought relevant legal and urbanistic innovations, which will serve to implement a new model for cities, including in the field that this study has focused on, environmental issues. The concern with environmental balance is clear.

3.3 THE REAFFIRMATION OF MASTER PLANS IN THE CITY STATUTE

In the City Statute Law, articles 39 to 42 deal with the Master Plan. As article 2, paragraph 4 of the City Statute Law states, the Master Plan must cover the entire municipal territory, as it is the basic instrument of urban development and expansion policy and an integral part of the planning process (GASPARINI 2002, p. 195).

With regard to the conceptualization of the Master Plan, Gasparini (2002) points out that it is a plan because it establishes objectives, deadlines and activities and those responsible for carrying out the actions, and it is a director because it establishes guidelines and principles for municipal urban development. Maricato (2001, p. 116) states that "the use of the legal instruments provided for in the City Statute and the Constitution of 88 is subordinate to the Master Plan".

The author (2001, p. 117-118) summarizes that

The Master Plan, as required by the City Statute, must overcome 1) the traditional mismatch between law and management and provide for management or the operational sphere;

2) the orientation of investments defined by private interests. We should therefore suggest orienting investments according to the public interest (social and environmental); 3) discriminatory, corrupt inspections restricted to the official city. We should propose an inspection standard; 4) urban planning regulations that apply to just one part of the city. We need to propose regulations that are citizen-based and universal; 5) technocratic and arrogant jargon. It must therefore be understood by the population so that it can be incorporated into the debate.

The author suggests a different approach to the Master Plan, as she sees it as an ideological instrument, rather than a tool for guiding management and investment. And its great challenge is to seek inclusive planning that addresses the problems of housing, public transport and environmental sustainability.

According to Santos Jùnior and Montandon (2011), the 1988 Federal Constitution promoted the principle of the social function of the city and property in Brazilian society. It thus affirmed the leading role of municipalities in drawing up the Master Plan as the basic instrument of urban development and expansion policy, which is compulsory for municipalities with more than twenty thousand inhabitants.

The main objective of the Master Plan, according to Santos Jûnior and Montandon (2011, p. 14), is to

[...] define the social function of the city and urban property in order to guarantee access to urbanized and regularized land for all social segments, to guarantee the right to housing and urban services for all citizens, as well as to implement a policy of democratic and participatory management [...].

The authors are well aware that municipalities have many difficulties in implementing their Master Plans, as most of them lack human, technological and material resources. Not to mention the lack of participatory councils to build and implement urban development policies.

However, these difficulties do not exempt municipalities from the responsibility of drawing up their plans, which must be approved by the City Council and in execution of the urban policy set out in article 182 of the Federal Constitution. In this regard, Gasparini (2002, p. 197) emphasizes that "within the municipality, the responsibility for implementing the master plan lies with the Executive, which is, in principle, more technically equipped, more knowledgeable about the local reality and closer to the wishes of the community".

Gasparini (2002) also adds that the Master Plan has a minimum content set out in the articles of the City Statute: in art. 42, item I, there is the delimitation of urban areas where delimitation and compulsory parceling, building or use will be applied, considering the existence of infrastructure and demand for use; in arts. 25 and 27, there is the right to build and the right to change the use of urban land. Articles 25 and 27 deal with the right to pre-emption; Articles 28 to 31 deal with the onerous granting of the right to build and the onerous alteration of the use of urban land; Articles 32 to 34 deal with consortium urban operations; and Article 35 establishes and regulates the transfer of the right to build.

In this case, according to the City Statute law, art. 40, § 1, "the master plan is an integral part of the municipal planning process, and the multiannual plan, the budget guidelines and the annual budget must incorporate the guidelines and priorities contained therein".

Since the Master Plan is an integral part of municipal planning, it is worth highlighting the relevance of this research in terms of evaluating the effectiveness and efficiency of this document. These discussions refer to Porto Nacional's Sustainable Development Master Plan, which, in Chapter 3, will evaluate the degree of effectiveness of environmental policies in order to promote the population's quality of life. The cartographic representation of this Master Plan can be seen on Map 2.

Map 2 - Porto Nacional Sustainable Development Master Plan

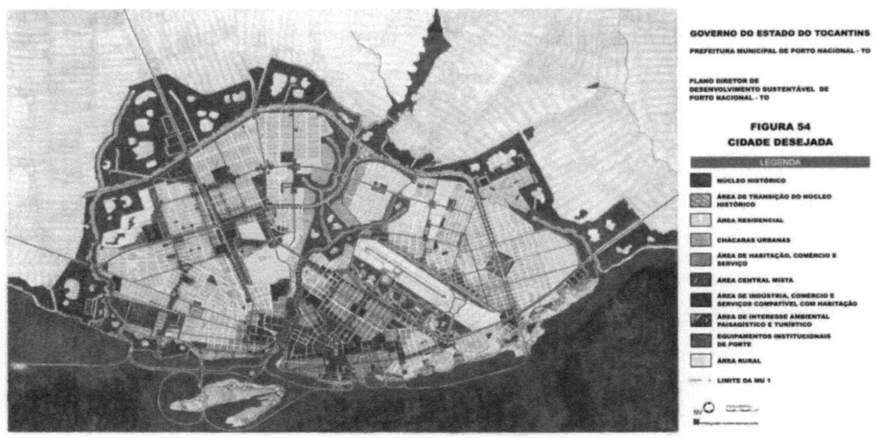

Source: Porto Nacional City Hall, 2006.

4. THE PDDSPN AND URBAN ENVIRONMENTAL INSTRUMENTS

This chapter aims to deal with the main instrument of Urban Planning and Management advocated in the City Statute: the Master Plan. It seeks to answer the specific objectives of this research: (i) to analyze the environmental instruments of the City Statute and their application in relation to Porto Nacional's Sustainable Development Master Plan and (ii) to analyze the interface/dialogue between environmental and urban policy instruments in the municipality of Porto Nacional in order to achieve the construction of urban environmental management in relation to environmental zoning.

According to Vitte and Keinert (2009), the Master Plan is an instrument for realizing the social function of the city when it meets the fundamental requirements of the City Statute law, so that it meets the needs of citizens and, consequently, promotes quality of life.

We can consider the Master Plan as an instrument capable of defining the rules of urban development in which society benefits in harmony with the environment, improving the quality of life for the inhabitants.

The Porto Nacional Sustainable Development Master Plan (PDDS-PN) is the basic instrument guiding this investigation. It sets out the guidelines that will lead to plans, programs and actions in the municipality. Despite being an instrument of urban policy, it is important to point out that the Municipal Master Plan Law does not have the self-applicability necessary for its instruments to make it possible to bring about urban reform as advocated in today's conception of urban planning, drawn up in Brazil in the City Statute with a marked reflection of the positions of the National Urban Reform Movement. However, through its guidelines, it guides the development of actions that promote sustainable urban development and the well-being of the population, through the amenity of the environment, which becomes a representative element of quality of life.

The Porto Nacional Sustainable Development Master Plan - PDDS- PN, regulated by Municipal Law 05/2006 and approved on 28/9/2006, was discussed by a multidisciplinary organizing committee made up of an Architect, Economist, Civil Engineer, Social Worker, Lawyer, Accountant, Geographer, Historian, Cartographer, all from various entities, such as: schools, universities, public security, municipal public administration, Catholic Church, SEBRAE and civil society. It was drawn up by the consultancy CA & CO - Camargo & Cordeiro Consultores Associados S/S Ltda.

This document is structured as follows in Table 2.

Table 2 - Structure of the PDDS-PN*

CHAPTERS	SECTIONS	SUB-SECTIONS	ARTICLES
I. MUNICIPAL POLICY FOR URBAN DEVELOPMENT AND EXPANSION	-	-	1 e 2
II. THE MUNICIPAL POLICY FOR URBAN DEVELOPMENT	-	-	3 e 4
III. CENTRAL OBJECTIVE AND STRATEGIC LINES	I. Economic development (arts. 8 and 9); II. Tourism Development (art. 10); III. Environmental Preservation (arts. 11 and 12); IV. Environmental Sanitation (art. 13); V. Electricity, Public Lighting and Communication (art. 14); VI. Urban Mobility (art. 15); VII. Public Safety (art. 16); VIII. Urban Space Structuring (arts. 17 and 18); IX. Housing (art. 19); X. Community Education and Health Facilities (arts. 20 and 21); XI. Municipal and Urban Management (arts. 22 and 23).	-	5 a 23
IV. land USE PLANNING	I. Macrozoning (articles 26 to 31)	-	24 a 42
	II. The Municipal Circulation System (arts. 32 to 34)	I. The Municipal Road System (arts. 35 to 36); II. The Urban Road System (arts. 37 to 39); III. Waterway traffic (art. 40).	
	III. Preservation of Historical and Cultural Heritage (arts. 41 and 42).	-	
V. URBAN POLICY INSTRUMENTS	I. Parcel or Building Compulsory, Time Progressive IPTU and Expropriation with Payment in Titles	-	43 a 62

	(arts. 46 and 47); II. Surface rights (arts. 48 and 49); III. Special Social Interest Zones (arts. 50 to 53); IV. Neighborhood Impact Study (arts. 54 to 58); V. The Granting of the Right to Build (arts. 59 to 62).		
VI. USE AND OCCUPATION OF URBAN LAND	-	-	63 a 68
VII. THE PLOTTING OF URBAN LAND	-	-	67 a 70
VIII. MUNICIPAL PLANNING	I. The Municipal Planning and Management System (arts. 73 to 83); II. The Municipal Information System (art. 84).	-	71 a 84
IX. FINAL AND TRANSITIONAL PROVISIONS	-	-	85 a 95

*Porto Nacional's Sustainable Development Master Plan.

Source: Oliveira (2009).

The scenario presented in Table 2, through the structuring of the PDDS-PN, comprises the Central Objective for the Municipality: the Strategic Lines and the corresponding action programs necessary to achieve the outlined objective, the urban planning proposal for the municipal seat, the guidelines for land use planning, the control of land use and occupation and the municipal environmental policy, within the principles of economic, socio-cultural, urban and environmental sustainability. It provides basic guidelines for structuring the Municipal and Urban Planning and Management System, which is necessary for implementing the Master Plan. It makes important considerations regarding the deployment of the process in new planning instruments, the Municipal Information System, the mechanisms for monitoring and evaluating the implementation of the Master Plan and popular participation and social control, in compliance with the guidelines established by the City Statute (OLIVEIRA, 2009).

The central objective of the Porto Nacional Master Plan, as mentioned in the introduction to this work, is to guide the urban expansion of Porto Nacional and recover its cultural and economic importance on the state stage. However, the Strategic Lines clearly refer to the macro-actions that must be carried out in order to achieve the central objective of the municipal development policy, and they explicitly include environmental actions, as shown in Article 6:

In order to achieve the central objective of the municipal development and urban expansion policy, the following strategic lines will be adopted: [...]II - rational and sustainable exploitation of the lake's economic and tourist potential. III - sustainable development of cultural and nature tourism. IV - protection of the environment; VII - qualification of the urban space, in order to eliminate socio-spatial segregation and valorize green areas;

§ 1ª The strategic lines provided for in this article will be implemented through specific action programs, projects and activities.

§ Paragraph 2 For the purposes of this Complementary Law, a green area is understood to mean spaces where there is a predominance of tree vegetation, encompassing squares, public gardens and urban parks, the central medians of avenues and the interchanges and traffic circles of public roads (PORTO NACIONAL, 2005).

The guidelines, in turn, provide the legal viability that is essential to the practice of the Strategic Lines, which are implemented through planning.

In this sense, it is up to this investigation to analyze the environmental instruments guided by the guidelines present in this Master Plan and their respective actions carried out within the urban macro-zoning:

- MUNICIPAL PLANNING INSTRUMENTS: MUNICIPAL policy and environmental zoning;

- LEGAL AND URBANITICAL INTRUMENTS: Neighborhood Impact Assessment;

- of THE LAND REGULARIZATION INSTRUMENTS: Special Social Interest Zone - ZEIS and Neighborhood Impact Study - EIA;

- THE DEMOCRATIZATION OF MANAGEMENT: Municipal councils and participatory budget management.

4.1 THE PDDS-PN MUNICIPAL PLANNING INSTRUMENTS AND ENVIRONMENTAL MANAGEMENT

Municipal planning takes precedence because it establishes the guiding principles for the municipality's development policy. In this sense, the PDDS-PN, according to Table 2, presents the Municipal Planning instrument, which establishes principles based on the Municipal Planning, Budget and Management system, which will be responsible for municipal urban planning actions to be carried out by the public authorities in partnership with the private sector and organized society. The preparation and execution of the municipal government's plans and programs must comply with the guidelines of the PDDS-PN, in which they will have permanent follow-up, monitoring and evaluation, in order to ensure their success and continuity.

The urban planning instruments laid down by the City Statute in Chapter II and Article 4 are present in the PDDS-PN Law, in Article 43, which includes the **multi-annual plan, budget guidelines and annual budget, sectoral plans, action programs and projects and the regulation of the parceling, use and occupation of urban land**. These instruments give the municipality responsibility for planning and budget execution, for drawing up actions that comply with the guidelines and for monitoring the use of urban land. In this sense, these guidelines condition the municipality to plan in a way that meets the real proposals of the Master Plan and guarantees the environmental quality of the municipality.

In order to analyze these real proposals, we will discuss the **Strategic Lines** that deal exclusively **with environmental preservation** presented in the PDDS-PN and their degree of effectiveness contemplated in the plans, programs and actions of each instrument that will be presented next. Thus, articles 11 and 12 of the PDDS-PN present:

Art. 11: Environmental preservation and the protection of ecosystems in balance with their sustainable use for the promotion of development shall be promoted by:

I - protection of water resources to ensure their perpetuation;

II   - enhancement of Permanent Preservation Areas and Conservation Units;

III   - protection of endemic or relevant ecological and environmental attributes;

IV   - recovery of water springs in the urban area of the municipal headquarters;

V   - promoting environmental education at all levels.

Sole Paragraph. The banks of flowing or dormant waters of springs shall be given non-building strips of environmental protection, at least 75m (seventy-five meters) wide, in addition to those determined by other competent legislation.

Art. 12: The municipal government shall formulate the Municipal Environmental Policy, in compliance with Art. 188 of the Organic Law of the Municipality, and draw up the Municipal Environmental Code (PORTO NACIONAL, 2006).

It is understood that the aforementioned guidelines should lead to actions in which environmental management is responsible for caring for, preserving and conserving the natural and built environment of the municipality of Porto Nacional.

4.1.1 Municipal Policy

As part of the Municipal Planning instrument, the Municipal Policy aims to: the full development of the social functions of the city and the guarantee of the well-being of its inhabitants; the participation of the respective community entities in the study, forwarding and solution of the problems, plans, programs and projects that concern them; the preservation, protection and recovery of the urban environment and cultural heritage; the creation and maintenance of areas of special historical, urban, environmental, tourist and public use interest; compliance with urban planning, safety, hygiene and quality of life standards; and restrictions on the use of areas at geological risk (PORTO NACIONAL, 2006).

The municipal policy is present in chapter II of the PDDS-PN, favoring the interpretation that it is the result of the Municipal Organic Law, approved on April 4, 1990, which ratifies, in article 188, the guideline that deals with the Environment and Natural Resources with the alteration of paragraph 3 through amendment 02/2011, as follows:

Art. 188. In order to ensure everyone's right to an ecologically balanced environment, it is the responsibility of the public authorities to propose and adopt a Municipal Environmental Policy.

§Paragraph 1. The Municipal Environmental Policy shall be guided by the provisions of this Organic Law and the following laws:

Building Code;

Municipal Zoning Law;

Land Use and Occupation Law;

Allotment Law;

Integrated Development Master Plan Law;

Law on the Protection of the Municipality's Historical, Cultural and Natural Heritage; Specific laws on environmental protection and preservation.

§Paragraph 5 - The public administration will develop the Municipal Environmental Policy, with the help of the Municipal Council for the Defense of the Environment, whose attributions and composition will be defined by law (PORTO NACIONAL, 1990).

Still discussing Municipal Policy with a focus on environmental management and as a result of Urban Planning, on December 22, 2006, the Municipal Legislature approved Law No. 1887/2006, which establishes the Municipal Environmental Policy Law of the Municipality of Porto de Nacional, with policies that advance in significant completeness for local environmental needs in order to guarantee environmental balance, materializing in the quality of life of the population.

The law in question is structured according to Table 3.

Table 3 - Structure of the Porto Nacional-TO Municipal Environmental Law

CHAPTERS	SECTIONS		ARTICLES
I. INITIAL provisions	I.	THE PRINCIPLES	1 a 5
	II.	OBJECTIVES	
	III.	INSTRUMENTS	
	IV.	GENERAL CONCEPTS	

Chapter	Section	Articles
II. THE MUNICIPAL ENVIRONMENTAL SYSTEM	I. THE STRUCTURAL ORGANIZATION OF THE SYSTEM II. THE MUNICIPAL environmental COUNCIL	6 a 16
III. MUNICIPAL ENVIRONMENT POLICY INSTRUMENTS	I. GENERAL STANDARDS II. ENVIRONMENTAL ZONING II. ENVIRONMENTAL EDUCATION III. THE CREATION AND MAINTENANCE OF SPECIFICALLY PROTECTED TERRITORIAL SPACES IV. ENVIRONMENTAL LICENSING V. ENVIRONMENTAL CONTROL AND MONITORING VI. ENVIRONMENTAL MONITORING VII. ENVIRONMENTAL RECOVERY VIII. THE MUNICIPAL ENVIRONMENT FUND ENVIRONMENT IX. SUSTAINABLE MANAGEMENT OF NATURAL RESOURCES X. scientific and technological development and its dissemination XI. ECONOMIC INSTRUMENTS XII. THE SUSTAINABLE DEVELOPMENT MASTER PLAN XIII. PROMOTING SOCIAL PARTICIPATION IN ENVIRONMENTAL ISSUES	17 a 77
IV. INFRACTIONS AND PENALTIES		78 a 97
V. FINAL TRANSITIONAL PROVISIONS		98 e 99

Source: Porto Nacional City Hall, 2006.

From this perspective, it is understood that the municipality has sufficient legislation that is explicitly presented and related to environmental aspects. Thus, these guidelines that support municipal policy can have a direct impact on environmental quality, based on the fulfillment of the social function of the city with the participation of society, leading to improved quality of life.

When it comes to evaluating this municipal policy based on the PDDS-PN, the municipality also has complementary municipal law no. 1883/06, in which article 5 defines "program" as an instrument for organizing government action, measured by indicators established in the Multi-Year Plan - PPA 2006-2009, and "project" as a set of operations, limited in time, resulting in a product. Finally, it defines an "activity" as a set of operations that is carried out continuously and permanently. This law establishes guidelines that outline priorities and goals in municipal administration, exemplified by article 2, item XIII, which deals with implementing environmental management programs.

Table 4 shows the municipal government's budget planning through the PPA - Multi-Year Plan 2006-2009/ 2010-2013, which seeks to comply with the regulations already discussed.

Table 4 - Municipal Government Work Programs published in 2006 and 2009

Specifications	Projects		Activities		Total	
YEAR	2007	2010	2007	2010	2007	2010
Environmental Management	34.000,00		867.000,00		901.000,00	29.000,00*
Coord. and Manut. Sec. de Medio Ambiente Development	0,00		805.000,00	1.875.300,00		
Environmental Preservation and Conservation	30.000,00		62.000,00			25.000,00*
Policy Management Environment	30.000,00		62.000,00			29.000,00*
Installation of the Library in the Ecological Park	10.000,00		0,00	6.000,00		
Promoting the shared construction of local Agenda 21	2.000,00	1.000,00	0,00			
Promoting Environmental Education and Responsibility	0,00		19.000,00	9.000,00		
Support a NGOs for the Preservation and Conservation of the Environment	0,00		2.000,00	1.000,00		

Environmental Control	4.000,00		0,00			4.000,00*
Environmental policy management	4.000,00		0,00			29.000,00*

*Values not defined as projects and activities, only as total expenditure. Source: Porto Nacional City Hall, 2006.

Law no. 1884/2006 provides for the updating of the Multi-Year Plan - PPA, for the period 2006/2009, of the municipality of Porto Nacional-TO, in which it outlines the actions for implementation in 2007, in the management of environmental policy, in accordance with the aforementioned article.

Law No. 1.999/2009, of December 29, 2009, provides for the 2010/2013 Multi-Year Plan (PPA) for the municipality of Porto Nacional-TO, and also deals with the implementation of environmental management programs with strategic objectives to be achieved. The actions planned include a set of activities to be carried out, among which we have chosen some that are the responsibility of environmental management and which are included in the PPAs already presented, as shown in Table 5.

Table 5 - Actions for the Environmental Management program in Porto Nacional-TO

ACTION	YEAR	degree of EFFECTIVENESS
a. Promote environmental education and responsibility, with a view to forming a culture of sustainable development in the municipality.	2013	Celebration of tree day in municipal schools.
b. Promoting environmental and urban quality through sanitation, management and control of urban space.	2013	Monthly report on the quantity per Kg of waste deposited in the Porto Nacional-TO controlled dump. Public hearing to present the Draft Municipal Water and Sewerage Plan.
c. Promote the shared construction of the local Agenda 21.		Unidentified actions.
d. Setting up the ecological library in the Ecological Park.		Unidentified actions.
e. Campaigning events for the preservation of the environment.	2013	Tocantins State Waste and Citizenship Forum.
f. Setting up a seedling nursery of cerrado species for the implementation and recovery of the urban park and the recovery of degraded areas.		Nursery with seedlings for gardening.

Source: Municipal Department of Housing and Environment of Porto Nacional-TO, 2006.

As part of the action to "Promote environmental education and responsibility, with a view to forming a culture for sustainable development in the municipality", actions were identified to commemorate the Day of the Tree, held on 19, 20 and 21/9/2013 and carried out through an Environmental Education Workshop and the planting of seedlings. The following municipal schools in Porto Nacional's Urban Macrozone participated in the action: Centro Municipal de Educaçao Infantil Tia Dedé, Centro Municipal de Educaçao Infantil Alice Maria, Centro Municipal de Educaçao Infantil Dona Aureny, Centro Municipal de Educaçao Infantil Ernestina Freire Aires, Centro Municipal de Educaçao Infantil Osvaldo Aires, Escola Municipal de Formaçao Integral Vereadora Marieta Pereira de Macedo, Escola Municipal Fany de Oliveira Macedo, Escola Municipal Padre Luso, Escola Municipal Delza da Paixao Pereira, Escola Municipal Uniao e Progresso, Escola Municipal Divino Espirito Santo, Escola Municipal Deasil da Silva, Escola Municipal de Formação Integrada Professora Generosa Pinto de Castro.

In line with the action to "Promote environmental and urban quality through sanitation, management and control of urban space", only a few monthly reports were identified on the weighing of the trucks that entered the Porto Nacional controlled dump. The weighing was divided into household waste, which was collected by the trucks under the supervision of the Urban Cleaning Directorate - DLU; the weight of rubble and branches collected by companies specializing in rubble and city hall trucks; and hospital waste, collected by a car specifically for this purpose, under the supervision of the DLU. The space was called a waste dump in the controlled dump of Porto Nacional-TO.

Also identified as an activity of the aforementioned action was the Public Hearing presented in the Official Gazette No. 3.990, of October 24, 2013, p. 75: a public notice calling on the community to attend on October 30, 2013 in the Plenary Hall of the City Hall, at 7:30 p.m., and in the District of Luzimangues, at the City Hall Subsidence Office, on October 31, 2013, at 7:30 p.m., with the purpose of discussing the Draft Municipal Water and Sewage Plan (PMAE).

In response to the action "Organize campaigns and events for the preservation of the environment", the Tocantins State Waste and Citizenship Forum - FELC- TO took place, with actions such as: solid waste management with the preparation of the Integrated Urban Solid Waste Management Plan, support for selective waste collection with the Association of Waste Pickers and the preparation and submission of projects to the Federal and State Governments. The forms of participation in the FELC were through organizing meetings/debates on the subject, publicizing the purposes of the FELC and supporting the waste pickers.

And in the action for "Setting up a nursery of seedlings of cerrado species for the implementation and recovery of the urban park and recovery of degraded areas", a nursery was found, according to Photos 6 and 7, located near the left bank of the Sao Joao stream, in Urban Macrozone 1, in the Jardim Querido sector. According to the information provided by the gardener in charge, this nursery has six (6) workers responsible for growing the plants, fertilizing the seedlings and organizing the space for the seedlings. Most of these seedlings are from gardening, but there are some Aroeira seeds that haven't germinated yet. In addition to gardening seedlings, only Pau Brasil, Cupuaçù and Pitomba seedlings were found.

Although this action has been carried out, it is still considered to be very fragile due to the size of the beds, the quantity and the small diversity of seedlings germinated.

Photo 6 - Seedling nursery for gardening

Source: GUILHERME, O. D. S

Photo 7 - Seedling nursery for gardening

Source: GUILHERME, O. D. S

4.1.2 Environmental zoning

As stated in the introduction to this research, the PDDS-PN establishes that macro-zoning is one of the instruments of land-use planning. Article 26 of the Plan states that the division into macro-zones aims to promote land-use planning, as well as the planning and proper implementation of the strategic lines and action programs defined by the Porto Nacional Sustainable Development Master Plan. Article 27 establishes the macro-zones as follows: Urban Macrozones (MU), Environmental Protection Macrozone (MA) and Rural Macrozone (MR):

Art. 28: Urban Macrozones are areas effectively designed to concentrate urban functions with the aim of:

I - optimize the urban and community facilities installed;

II - guide the urban expansion process;

III - to condition urban growth to the capacity of urban and community facilities.

§   1 Under the terms established in the *caput of* this article, the following are Macrozones

Urban:

I   -the seat of   the municipality, as Urban Macrozone 1 - MU 1;

II   -District of   Luzimangues, as Urban Macrozone 2 - MU 2;

III   - Porto Nacional Agro-Industrial District, created by Municipal Law No. 1.308 of August 12, 1991, amended by Law No. 1.305 of June 12, 1992, as Macrozone 3 - MU 3;

IV   - Escola Brasil village, as Macrozone 4 - MU 4;

V   - village of Nova Pinheirópolis, as Macrozone 5 - MU 5 (PORTO NACIONAL, 2006).

The Environmental Protection Macrozone - MA dedicated to the protection of ecosystems and natural resources is made up of the Environmental Protection Area - APA do Lago de Palmas, and the Rural Macrozone - MR is made up of the rest of the municipal territory, destined for agricultural, mineral extraction and agro-industrial activities.

Map 3 - Urban Macrozone 1 of Porto Nacional-TO

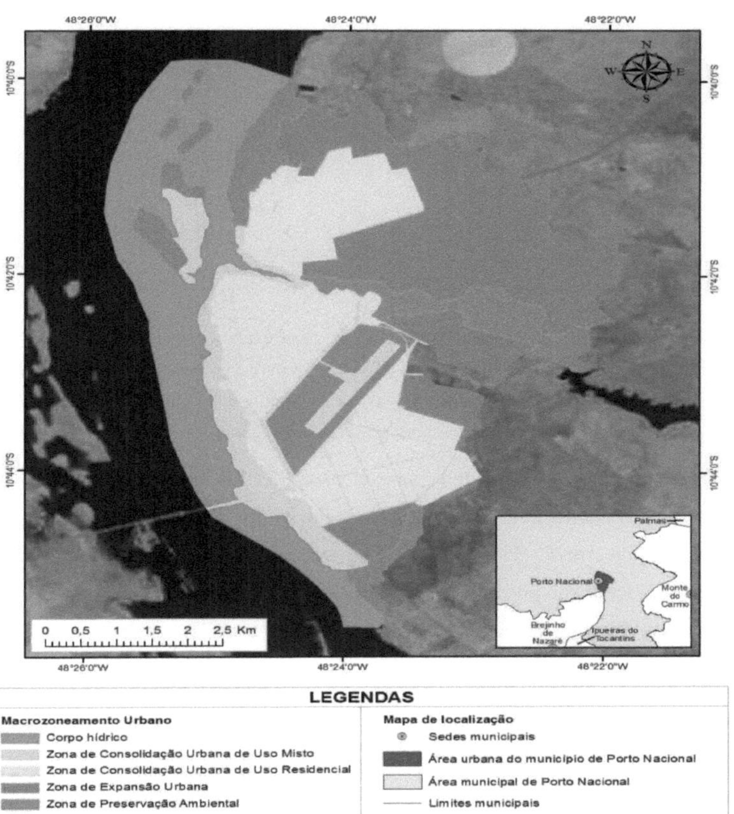

Organization: MORAES, Erton Inacio M. de

Map 4 represents the Environmental Protection Macrozone (MA).

Map 4 - Macrozoning map of the municipality of Porto Nacional-TO

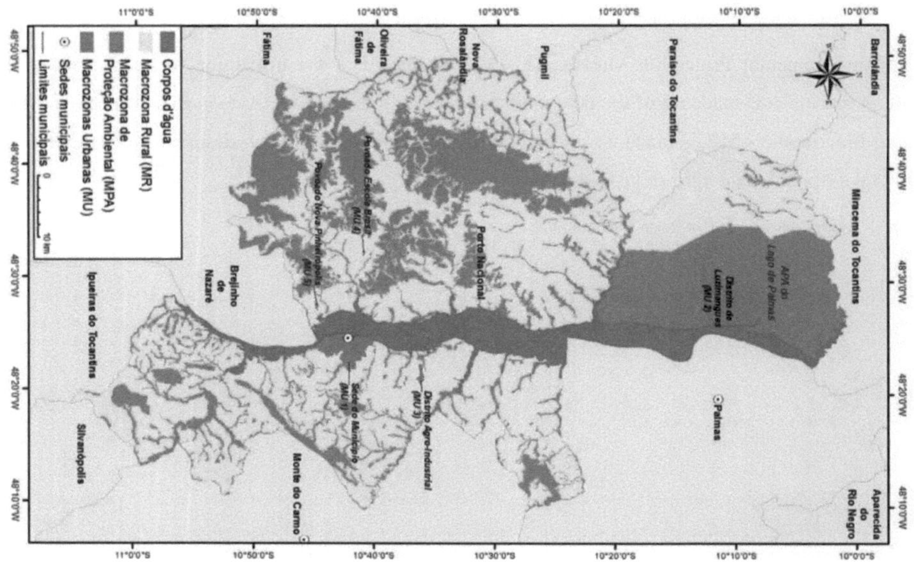

Organization: MORAES, Erton Inacio M. de. As shown on Map 4, the Environmental Protection macro-zoning covers a large part of the municipality, with the APA of Lago de Palmas in the north, which is approximately 50% (fifty percent) Environmental Protection Area. Throughout the municipality, there are several areas designated as APAs. It is understood that these areas are, in particular, springs and the surroundings of water sources.

The evaluation of environmental management actions is based on the question: how has the municipality been treating this Macrozone? How does the municipality ensure the preservation of these Environmental Protection Areas? What projects, programs and actions has the municipality planned and carried out in order to comply with the guidelines of the Master Plan that guarantee the delimitation of these areas and their preservation?

The information provided by the Municipal Department of Housing and the Environment was that there are no records of actions from previous administrations. As such, only the actions reported in the implementation of environmental management by the municipality of Porto Nacional are included.

ACTION 1 - The creation of Law No. 1887 of December 22, 2006, which establishes the Municipal Environmental Policy Law for the Municipality of Porto Nacional. It contains articles dealing exclusively with environmental zoning: (i) article 4, environmental zoning is one of the instruments of environmental policy; (ii) article 12 and item III, the Municipal Environmental Council must establish basic and well-founded criteria for drawing up environmental zoning, and may or may not approve the proposal submitted by the competent municipal body; (iii) Article 19, Environmental Zoning consists of defining areas of the municipality's territory in order to regulate activities and define actions to protect and improve the quality of the environment, taking into account the characteristics or attributes of the areas; (iv) Article 19, second paragraph, states that Environmental Zoning should be used to prepare land use zoning, specifically for the municipality's headquarters; (v) Chapter IV states that the Municipal Department for the Environment and Sustainable Development will impose penalties (fines) for violations of this Law.

However, the municipality lacks actions under the Municipal Law that are effectively applied. One example is the creation of the Municipal Environment Council - CMMA, suggested in Municipal Law 1887/2006 and created by Law 2015 of October 8, 2010. For the current administration, this council does not exist, justifying that it is made up of 17 (seventeen) members and among them *there are* 4 (four) municipal secretaries. Since the transition of municipal managers took place in 2013, they have not yet been able to organize the Council. In the meantime, some of the supervisory duties that would be the council's responsibility are not being carried out, so individual and collective social values and the conservation of the environment, which is essential to a healthy quality of life, are at the mercy of their vulnerability.

ACTION 2 - The creation of Law 2.132 of December 5, 2013, which provides for the revitalization of the Ribeirao Sao Joao in the Municipality of Porto Nacional. It contains articles such as the 6th, which states:

The Porto Nacional City Hall, through the Environment Department, will clean and remove the sediment deposited at the bottom of the river, as well as permanently monitoring the degree of silting, in order to promote its prompt correction (PORTO NACIONAL, 2013).

To date (March 2014), Porto Nacional's Municipal Housing and Environment Department has not been informed of this law. The Sao Joao stream, which crosses Urban Macrozone 1 in an east-northwest direction and flows into the Tocantins river lake, which is located to the west of Urban Macrozone 1, has a lot of debris, such as plastic, furniture wood, mattress foam, aluminum cans and various inorganic materials, as well as plant debris that is thrown into the stream or on its banks less than a meter from the bed. It can be seen that during the flood period this debris is very noticeable on the banks, as can be seen in Photos 8, 9, 10 and 11.

Photo 8 - Rubbish on the left bank of the R. Sao Joao

Source: GUILHERME, O. D. S

Photo 9 - Rubbish and debris on the left bank of the R. Sao Joao

Source: GUILHERME, O. D. S

Photo 10 - Rubble on the left bank of the R. Sao Joao

Source: GUILHERME, O. D. S

Photo 11 - Rubbish and debris on the left bank of the R. Sao Joao

Source: GUILHERME, O. D. S

On the right bank of the Sao Joao stream there is a steep slope that runs along the whole of Urban Macrozone 1 until it flows into the Tocantins River, now the lake of the Eduardo Magalhaes Hydroelectric Power Station. During the rainy season, the torrents bring a lot of sediment to the stream bed due to the steep terrain and also because the area is surrounded by several housing developments, such as Jardim América, Beira Rio and Vila Nova.

According to local residents, the Sao Joao stream has an irreparable social value for them, because people used to fish there, wash clothes and even collect water for domestic use. Today, however, the waters of this stream are completely unusable.

ACTION 3 - The creation of Complementary Law no. 07/2006, which deals with the Urban Land Parceling of the Municipality of Porto Nacional and originated from the PDDS-PN. Article 70 legislates that: "Any urban land parceling in the Municipality will have to be approved by the City Hall, under the terms of the federal, state and municipal laws on urban land parceling and the environment" (PORTO NACIONAL, 2006). And article 29 of this law states that "the utilization coefficients of the Urban Macrozones will be defined in the Complementary Law on Land Use and Occupation in the Urban Macrozones of the Municipality of Porto Nacional" (PORTO NACIONAL, 2006).

However, what we found was a great demand for allotments that follow the urban macro-zoning, stretched out by complementary laws, in an indiscriminate manner, without consulting the community and much less the councils that should be created to monitor and even discuss the use and occupation of urban land. As an example, we have Law 07/2006 on Urban Land Parceling, which establishes in article 12 that a plot must have a minimum of 360m^2 (three hundred and sixty square meters), yet the recently parceled plots have an average of 180m2 to 260m2.

The Department of Housing and the Environment of the municipality of Porto Nacional reported that the municipality's Environmental Protection Macrozone is unknown to them and that to date there is no formalized document (maps) with a cut-out of these protection areas. Nor have they identified any projects, programs or actions with a more efficient content that would meet the PDDS-PN guidelines for guaranteeing these preservation areas.

4.2 THE LEGAL AND URBAN PLANNING INSTRUMENTS OF THE PDDS-PN AND ENVIRONMENTAL MANAGEMENT

The legal and urban planning instruments drawn up by the PDDS-PN are treated in the City Statute Law, article 4, item V, as legal and political institutes. The terms are synonymous when one looks at the specifications of each. For the Porto Nacional Master Plan, the legal and urban planning instruments are: a) administrative servitude; b) time-progressive urban property tax - IPTU progressivo; c) tombamento; d) expropriation; e) compulsory parceling or building; f) expropriation with payment of titles; g) special usucapiào of urban property; h) right of superficie; i) **neighborhood impact study - EIV**; j) onerous granting of the right to build.

In the City Statute law, the **Neighborhood Impact Study (EIV)** is an independent instrument, according to item VI of article 4, while in the PDDs-PN, article 43 and item II, the EIV is part of the **legal and urbanistic instruments**, as an instrument of Urban Policy, together with the instruments mentioned above.

In this sense, the **neighborhood impact study (EIV)** was chosen for this investigation because it adds to the evaluation of environmental management in the municipality of Porto Nacional. This is because the Neighborhood Impact Study (NIS) is undoubtedly a major contribution of the City Statute in terms of urban environmental management. It is an instrument that aims to mitigate conflicts over the use and occupation of land, creating a new possibility of intermediation between the interests of urban entrepreneurs and the population directly impacted, in order to safeguard minimum standards of quality of life.

4.2.1 Neighborhood impact study

The EIV is an instrument for mitigating the environmental impacts caused by a project or activity that may directly or indirectly affect natural resources and society. Articles 54 to 58 of the PDDs-PN were drawn up to ensure the safety of both society and the environment:

Art. 54: The neighborhood impact study shall be carried out in such a way as to consider the positive and negative effects of the enterprise or activity on the quality of life of the population living in the area and its surroundings, including the analysis of at least the following issues:

I   -population densification;

II   -urban and community facilities;

III   -land use   and occupation;

IV   -   real estate appreciation;

V   - traffic generation and demand for public transport;

VI   -   ventilation and lighting;

VII   - urban landscape and natural and cultural heritage;

VIII   - solid waste generation;

IX   - socio-economic impact on the population living or working in the area;

X   - noise and visual pollution.

Art. 56: In order to eliminate or minimize the negative impacts to be generated by the project, the Municipal Executive Branch shall request, as a condition for project approval, alterations and additions to the project, as well as improvements to the urban infrastructure and community facilities, such as: [....] V - maintenance of buildings, façades or other architectural or natural elements considered to be of landscape, historical, artistic or cultural interest, as well as environmental recovery of the area; VII - percentage of social housing in the development; IX - maintenance of green areas.

Art. 57 - The preparation of the EIV does not replace the environmental licensing required under the terms of environmental legislation.

Art. 58: The documents comprising the EIV/RIV shall be made public and shall be available for consultation by any interested party at the competent municipal body.

Sole paragraph. Copies of the EIV/RIV will be provided when requested by residents of the affected area or their associations (PORTO NACIONAL, 2006).

According to Araujo (2003), the EIV should analyze issues such as population density, overload on urban infrastructure, traffic generation and demand for public transport, and possible damage to the urban landscape. Of course, these issues can also be understood as environmental, since they refer to the built environment, but they are first and foremost an urbanistic concern.

Prieto (2006, p. 11) states that "the study instruments - EIA and EIV - if properly applied by municipalities, can guarantee the urban population that the environmental balance will be safeguarded".

Thus, of the **legal and urban planning instruments** present in the PDDS-PN, the EIV is the one of interest in this investigation in order to diagnose its degree of effectiveness, since this Master Plan presents guidelines established for the environment, defining specific instruments aimed at environmental sustainability.

In this sense, the EIV and its respective Neighborhood Impact Report -

RIV were established in Porto's Sustainable Development Master Plan

However, a specific rule will define the projects and activities in urban areas that will require the preparation of the EIV and RIV in order to obtain licenses or authorizations for construction, expansion or operation. However, as of this writing, we have not been able to verify its existence, as the technicians from the municipality's Housing and Environment Department claim to be unaware of this instrument.

4.3 THE PDDS-PN LAND REGULARIZATION INSTRUMENTS AND ENVIRONMENTAL MANAGEMENT

Listed in the PDDS-PN as instruments for implementing sustainable development policy are the land regularization instruments: concessions of use and real right of use, special concessions of use for housing purposes, special zones of social interest - ZEIS and environmental impact studies - EIA. For the purposes of this study, only special social interest zones (ZEIs) and environmental impact studies (EIAs) will be considered. According to Rolnik (1994), the purpose of the ZEIs is to identify irregular settlements in order to promote land regularization, and they can be an important instrument for including the population. And the EIA, according to Ribeiro (2004), is a document in which the environmental assessment information is substantiated, which presents and discusses impacts considered relevant and proposes mitigating measures and a monitoring plan. Ribeiro (2004) also argues that community participation is an extremely important requirement, according to the CoNAMA resolution of 1987, whereby the community can give its opinion, suggest, disagree and hold public hearings, but it is not a decisive factor, although it must take place as soon as possible so as not to cause major problems once the project is underway.

We will deal with the Special Zone of Social Interest (ZEIs) and the Environmental Impact Study (EIA) in relation to the PDDs-PN and municipal environmental management for the aforementioned diagnosis of the degree of application.

4.3.1 Special Zone of Social Interest - ZEIS

The ZEIS proposal drawn up by the City Council sought to take into account the production and maintenance of social housing and the regularization of public and private land that is already occupied, especially by low-income populations. Three types of zones were considered, as can be seen in the following quote from article 51 of the PDDS-PN:

I.   ZEIS1 : located in the   Sao Francisco subdivision;

II. ZEIS2 : located in the   Sao Vicente   allotment;

III. ZEIS3:   the entire   Urban   Macrozone 2 - MU2 (Luzimangues   district).

According to Coriolano (2011), it is in the market's interest to establish affordable housing in these areas and the state encourages this to happen, including the provision of public investment to regularize the situation of allotments set up without infrastructure and, consequently, the appreciation of the urban voids that are located between these areas and the central areas.

Porto Nacional's Master Plan establishes guidelines that assign the Municipal Executive to decree, in compliance with the land occupation and parceling law, the delimitation of areas to be integrated into the ZEIS and to draw up intervention programs in accordance with federal legislation. The Plan also states that the ZEIS will have specific urban planning and land regularization plans, in order to implement the guidelines of articles 52 and 53 of the PDDS-PN:

I - to adapt property and its social function, prioritizing the right to housing over the right to property;

II   - effectively control the use and occupation of land;

III   - allocate public investment to meeting local needs, especially housing, urban and community facilities, roads and the environment;

IV   - create instruments that restrict real estate speculation and prevent the indirect eviction of residents;

V   - encouraging and guaranteeing community participation, as well as that of organized civil society entities, in the process of regularizing funding and urbanizing areas;

VI   - implement the installation of urban and community facilities, in line with the needs and socio-economic and cultural characteristics of the residents of the ZEIS;

VII   - prioritize the use of local labor;

VIII   - preserving and strengthening existing productive activities in the area;

IX   - drawing up specific urban planning and building regulations.

Art. 53: The regulation of the ZEIS must include, for each of them, an Urbanization and Legalization Commission, which will be responsible for:

I - coordinate and supervise the preparation and execution of the Urban Planning and Legal Regularization Plan for the respective ZEIS;

II - to mediate matters of interest to the ZEIS with the bodies of the direct or indirect administration;

III   - draw up quarterly reports on the progress of the specific Urban Planning and Legal Regularization Plan;

IV   - drawing up a register of people to be removed to plots or houses included in the specific project, in accordance with priority criteria established between the Municipal Executive and the community;

V   - settle issues not covered by this Complementary Law, as well as doubts arising from its application, with regard to the specific project;

VI   - monitor the use of allocated budgetary and financial resources;

VII   - draw up a closing document for the Specific Plan which, when submitted to the Mayor, will extinguish the Urbanization and Legalization Commission.

Sole paragraph. Each Urbanization and Legalization Commission will be made up of representatives from the Municipal Government, the residents' representative entity, the technical sector and the Municipal Legislative Branch (PORTO NACIONAL, 2006).

It is important to emphasize that the PDDS-PN is an important instrument for the inclusion of the population, based on the effectiveness of the application of its guidelines by the municipal executive.

However, as a diagnosis of the intervention programs in the areas delimited as ZEIS, not even the Special Zones were identified in any of the documents that may circulate in the municipality's Social Development Secretariat and other municipal secretariats. These neighborhoods can be identified within urban macro-zoning 1 in Porto Nacional, but their socio-spatial segregation is visible when these spaces are separated from the others by an extensive strip of unbuilt lots surrounded by closed vegetation, as is the case with ZEIS 1: located in the Sao Francisco subdivision. To the west of the Francisco sector is the Vila Operaria sector, built by low-income housing, which is divided by just one street, as shown in photos 12 and 13.

Photo 12 - Street separating the Vila Operària sector and ZEIS 1 - Sao Francisco

Source: GUILHERME, O. D. S

Photo 13 - Street in ZEIS 1 Sao Francisco

Source: GUILHERME, O. D. S

ZEIS 2, located in the Sao Vicente housing estate, is also spatially separated by an area of urban

73

voids, and next to this sector is a new housing estate, as shown in Photos 14 and 15.

Photo 14 - Affordable housing to the west of the Sao Vicente sector

Source: GUILHERME, O. D. S

Photo 15 - Urban voids at the entrance to the Sao Vicente sector.

Source: GUILHERME, O. D. S

ZEIS 3 in Urban Macrozone 2 - MU2 (Luzimangues district) has undergone countless transformations in terms of land parceling since the creation of the Porto Nacional Master Plan Law in 2006. According to Pinto (2012), urban allotments have been set up in Luzimangues and there is a significant disparity between the number of inhabitants and the number of plots sold. It is clear that the process of rural/urban transformation is taking place effectively, and that what was defined through legal aspects and regulations is now reflected concretely in the territory.

Photo 16 - Luzimangues subdivision

Source: OLIVEIRA, Josimar

Nor were any intervention programs identified, nor any recognition of these areas as a Special Zone of Social Interest in order to comply with the aforementioned guidelines of the Master Plan. These neighborhoods continue to be totally abandoned, and all that could be identified was a school and the construction of a Health Unit in the Sao Francisco sector. The Sao Vicente sector is totally abandoned, and the streets are unpaved and there is precarious infrastructure.

4.3.2 Environmental Impact Assessment - EIA

The instrument presented above is only mentioned in the body of the PDDS-PN Law, and no strategy has been presented for its implementation. In Chapter V, on Urban Policy Instruments, the Master Plan Law establishes other instruments, some of which appear in more detail. However, when it comes to land regularization instruments, only the ZEIS appear in detail. With regard to the instrument in question, it is worth analyzing the treatment given to it in the creation of complementary laws that will contribute to environmental actions.

The PDDS-PN, in article 57, emphasizes that the EIV should not be confused with the Preliminary Environmental Impact Study (EIA), a requirement of the environmental license required by Law n. 6.938, of 1981 (National Environmental Policy Law). Article 38 of the City Statute itself states that the preparation of the EIV does not replace the preparation and approval of the EIA, required under

75

the terms of environmental legislation.

Coriolano (2011), in his approach to EIA and EIV, states that the EIA will most often be analyzed by SISNAMA's state body, since the National Environmental Policy Law establishes that environmental licensing is, as a rule, the responsibility of SISNAMA's state body. In the case of a project with a regional or national impact, it will be analyzed by the federal agency. The EIV will - always - be analyzed by a municipal body. For the same author, there are serious disagreements about the legal basis for environmental licensing carried out by the SISNAMA municipal body, provided for in CONAMA Resolution 237 of 1997, since art. 10 of the National Environmental Policy Law does not provide for the possibility of the municipal body acting as licensor. For her, the municipality can, by its own law, impose an environmental licensing process on certain undertakings and, in cases of potentially significant impact, require an EIA. When environmental licensing is carried out at municipal level, the boundaries between the EIA and the EIV are not so clear. It will be up to the municipal law to spell them out.

In these terms, no intervening action by the municipal executive through the Environment Department was identified with regard to the Environmental Impact Study. The EIA is considered by those responsible for the municipality's Environment Department to be an unknown element for them.

In this circumstance, we can identify the disregard of the Porto Nacional Municipal Executive for the application of its Master Plan Law, the irreverence towards the environment and the omission of commitment to social development promoted by environmental actions supported by national, state and municipal guidelines.

4.4 THE INSTRUMENTS FOR DEMOCRATIZING THE MANAGEMENT OF THE PDDS-PN and environmental management

The instruments for democratizing urban management are extremely important given the process of intense urbanization that Brazil is experiencing. Here, social stratification and inequality prevail in terms of full access to quality urban life, due to the lack of basic equipment to maintain its existence.

Discussing municipal environmental planning and management from a socio-environmental perspective means adding human needs to the environment, so that the urban space ceases to be a hostile, unwelcoming and strange place and the city is rescued as an elective place in the civilizing process. This place must promote new meanings with regard to quality of life, through the population's desires, dreams, fears, possibilities to choose and decide, in short, democracy, the strengthening of citizenship and the right to the city.

In this sense, we sought to discuss the legislation that guides the implementation of urban democratization. According to article 43 of the City Statute Law, the following instruments, among others, should be used to guarantee the democratic management of the city: (i) collegiate bodies for urban policy, at national, state and municipal level, (ii) public debates, hearings and consultations, (iii) conferences on matters of urban interest, at national, state and municipal level and (iv) popular initiative for bills and urban development plans, programs and projects. Thus, Vitte (2009, p. 119) argues that it is necessary to redefine the concepts and meanings of democracy and citizenship and insert that of sustainable development. This makes it possible to redefine the matrices of rationality and adopt the principle of governance as a social construction that emerges from social movements, socio-environmental conflicts and is constituted from the various dialogues between local knowledge. Thus, governance

[...] it must be constituted within a new rationality, the environmental one, in which sustainability is broad and encompasses the pair of society and nature, but which allows us to think about the principle of identity, but also the identity and plurality of the other. [It also allows us to rethink collective identities, cultural diversity and the social reappropriation of nature (VITTE, 2009, p. 119).

It can be seen that the author considers the democratic game of building environmental rationality and quality of life to be a space of complexity, since it involves the construction of identity, tolerance, solidarity and new links between culture and nature that are maintained through the dialog of knowledge and the construction of citizenship. We can affirm that the participation of society as an act of democracy in government decisions, as well as being a legal act, is rational.

With regard to the Participatory Management of the Municipality of Porto Nacional, article 22 of the PDDS states that in order to develop the planning and management capacity of the Municipal Administration, as determined by the City Statute, the municipal and urban planning and management process needs to be implemented in a participatory manner.

And article 72 reaffirms that,

[...] in accordance with what is established in the City Statute, the preparation and execution of the plans and programs of the Municipal Government will obey the guidelines of the Porto Nacional Sustainable Development Master Plan and will have permanent follow-up, monitoring and evaluation, in order to ensure their success and continuity (PORTO NACIONAL, 2006).

This instrument for democratizing management is subdivided into municipal councils and participatory budget management. In this sense, the Master Plan law ratifies Municipal and Urban Management in articles 22 and 23:

Art. 22: The development of the planning and management capacity of the Municipal Administration in order to implement the municipal and urban planning and management process in a participatory manner, as determined by the City Statute, will be carried out through:

I - strengthening the planning and democratic management capacity of the municipality and the city;

II   - adapting the administrative structure to the functions required by the implementation of the Master Plan;

III   - training the technical teams of the different areas of the Municipal Administration;

IV   - integration of actions between the different bodies to rationalize the use of resources and maximize results;

V   - coordination with other spheres of government for development actions;

VI   - promoting partnerships, decentralization and the convergence of actions.

Art. 23: The development of forms of participation by the private sector, the third sector and the different segments of civil society in the municipal and urban planning and management process, to be established as soon as the Sustainable Development Master Plan is approved, will take place through:

I   - stimulating the creation of organizations representing society;

II   - promoting training for participation;

III   - using different channels of communication with the population.

To this end, its Sustainable Development Master Plan is a fundamental instrument for guaranteeing social participation in order to boost quality of life, through plans, programs and projects that allow its guidelines to be applied.

Examples of community participation in environmental activities that promote the well-being of the population, as mentioned above, include the State Waste and Citizenship Forum:

I.   The Regional Environmental Conference was held in Porto Nacional in May 2013. It was attended by the community, the municipal executive, the state and municipal environmental secretaries and the public prosecutor's office;

II.   the Environment Week, from June 3 to 7, 2013, which included the distribution of Pau Brasil seeds, tree planting, lectures and the handing out of educational pamphlets in the streets of Porto Nacional.

In this research, we had access to the budget execution from 2009 to 2013, made available by the Planning Department of the Porto City Hall.

National, as shown in Table 6.

Table 6 - Amounts executed by the Environment and Development Secretariat

2009	2010	2011	2012
R$ 1. 969.813,49	R$ 2. 394.896,91	R$ 2.664.553,76	R$ 4.466.360,34

Source: Porto Nacional City Hall Planning Department, 2006.

The figures shown in Table 6 refer to the rendering of accounts by the municipal executive regarding the use of environmental responsibility resources. The total expenditure for the Environment and Development Department in 2009 was R$1,969,813.49 (one million nine hundred and sixty-nine thousand eight hundred and thirteen reais and forty-nine cents). However, no action was identified to justify these expenses, except for employer's obligations, consumables, outsourced services from individuals and companies, and permanent equipment and materials.

The same happened in 2010, but with a higher sum of R$2,394,896.91 (two million, three hundred and eighty-four thousand, eight hundred and ninety-six reais and ninety-one cents). And in 2011, the amount was R$2,664,553.76, with no descriptive or specific actions relating to environmental management, which leads to a weak interpretation of social responsibility in the use of public resources. In 2012, the total expenditure for this environmental secretariat was R$4,466,360.34 (four million, four hundred and sixty-six thousand, three hundred and sixty reais and thirty-four cents).

In this sense, society is at the mercy of this information and the true destination of these funds, which are made available to public managers every year with the aim of contributing to the conservation and preservation of the environment in a citizen-friendly and democratic manner. There are no significant events held by the municipal administration with the participation of society in the planning, elaboration, execution and evaluation of projects and programs that guarantee the right to a decent city and a healthy environment.

5.   FINAL CONSIDERATIONS

The main question we sought to answer in this work was whether the guidelines of the Porto Nacional Sustainable Development Master Plan have been applied as environmental management tools in the urban macro-area of the municipality of Porto Nacional, in order to promote the quality of life of the population.

To reach this conclusion, discussions began with a macro view of the subject. To this end, we sought a debate with a historical and conceptual approach to Urbanization, Urbanism and Urban Planning. Urbanization is a process that originated in cities that preexisted approximately 3500 years BC, while urbanism appears as an action to deal with the problems that urbanization has promoted and, in this context, urban planning is necessary in order to prevent the problems promoted by urbanization and to make urbanism effective. However, this is not the Brazilian reality. The process of urbanization is disordered and has caused numerous economic, social, environmental and other problems, directly affecting the quality of life of the population. Urban planning has also failed to meet real needs due to a lack of urban planning that deliberates management based on the laws that govern cities, especially the City Statute.

We also discussed the political and legal instruments of urban planning in Brazil, at which point we had the opportunity to understand the genesis of the City Statute Law, which emerged with the 1988 Federal Constitution, through its articles 182 and 183. The City Statute was created in 2001 and is understood as a set of general, specific and management guidelines and instruments for the implementation of urban policy, seeking to make the Master Plan compulsory as part of the municipal planning process. And, if executed as a priority, it is an instrument par excellence for implementing urban policy. In the specific case of Porto Nacional, this law has been complied with up to the point of drawing up the Master Plan and its complementary laws, but the municipal administration has only considered the drafting and approval of bills to be "mission accomplished". What was actually seen was a real lack of knowledge of these laws on the part of the managers responsible for implementing them.

The diagnosis obtained by evaluating the application of the environmental instruments contained in the City Statute, as evidenced by the guidelines in the PDDS-PN, as part of municipal management, allows the following considerations to be made.

•   Porto Nacional's Sustainable Development Master Plan makes it possible to say that it is a well-designed document that complies with the rules established by the City Statute. However, when this analysis is transposed from the field of theory to the field of practice, it can be seen that a large part of what was planned ends up not being implemented or implemented with adaptations to momentary needs.

• This situation can be observed in the case of the municipal policy instrument and urban environmental management in Porto Nacional, as there is a Municipal Organic Law from 1990 which directs the municipality to adopt a municipal environmental policy. The Master Plan drawn up in 2006 stems from the municipal environmental law and there are other complementary laws, such as those on land use and occupation and land parceling. However, the results of an analysis of the municipality's budget guidelines reveal a lack of awareness of these laws when it comes to allocating resources for plans, programs and projects to implement these laws.

• According to the evaluation of the environmental zoning instrument, what was found was that not even a project dealing with the urban environmental macro-zone (still unknown to City Hall technicians) has come to fruition. Some vague intentions of action to revitalize the Ribeirao Sao Joao were found, but in view of the new housing developments that are approaching the banks of the Ribeirao, concern for the revitalization and preservation of this spring does not seem to be a priority of the municipal administration.

• As a legal and urban planning instrument of great importance, the EIV puts the guarantee of environmental balance in "check", due to the fact that it is unknown to municipal agents and does not present any impact report of any development installed in the municipality.

• Faced with the instruments of land regularization, both the EIA and the ZEIS explicitly represent the interests of the development policy that economic neoliberalism has implemented in the country. The managers of the municipality's environmental agency claim to be unaware of the EIA and we haven't found any document proving the existence of this study in the municipality, nor does the PDDS-PN legislate with much conviction on the EIA. The ZEIS exist as segregated areas and do not receive the attention provided for in the PDDS-PN.

• The most contradictory result found is that of budget execution for environmental management. It was observed that the amounts are in the order of millions and double every year, yet in their allocation there is no transparency of specific actions, nor does the population know or participate in this budget execution, a fact that contradicts articles 82 and 83 of the PDDS-PN.

The conclusion of the research is that the Master Plan is the instrument capable of defining the rules of urban development in which society benefits in harmony with the environment, improving the quality of life for all inhabitants.

However, urban planning and environmental management in Porto Nacional are comparable to a product of modern society and lack intelligent human activity that highlights the socio-environmental impacts of a better structured and executed planning system.

To this end, the discussions and analysis presented throughout this research confirm the proposition that the City Statute, through the implementation of the Master Plan, will enable territorial management that articulates environmental policies in the municipality. And the environmental policy instruments, once used by the municipal government, will have an impact on a more effective and environmentally friendly urban development policy. In this sense, we recommend

•	articulate the environmental guidelines contained in the PDDS-PN, the Environment Law and other complementary laws that refer to the urban environment with municipal policy such as the budget law (PPA and LDO) and, subsequently, with actions: plans, programs and activities that can be implemented;

•	make the CMMA deliberative and, through it, seek mechanisms for community participation in diagnoses, debates, complaints, suggestions and solutions on urban environmental issues;

•	seek to involve all sectors of human activity in environmental issues within the municipality, a situation that requires dialogue between the spheres of public administration and civil society;

•	structuring a local environmental management system that seeks an urban environmental policy within a broad and sustainable context, aimed at the quality of life of societies and the environment in which they live;

•	to form a strong normative base that will provide the basis for the actions that will follow from it, as well as integrating all the services linked to environmental and urban issues, not forgetting the means of information/communication and socio-environmental education, which together provide the basis for qualified popular participation in decision-making processes in the environment, seen from a holistic perspective. In a way that activates and strengthens oversight in the municipality and gives broad visibility to the Master Plan and builds mechanisms for popular participation.

•	to rethink the relationship between public power and the citizen, perhaps the spaces for participation are outdated and distant from the citizens who must necessarily feel that they belong to the reality that surrounds them. Social inclusion and participation programs are worthless if they don't generate a sense of belonging;

•	promoting environmental education in order to provide citizens with a critical and holistic view of the reality that surrounds them, making them feel they belong to the environment in which they live, permeating the issue of popular participation, which must be accompanied by knowledge and not just information.

6. REFERENCES

ABIKO, Alex Kenya; ALMEIDA, Marco Antonio Plàcido de; BARREIROS, Màrio Antônio Ferreira. **Urbanism**: History and Development. Polytechnic School of the University of São Paulo: Department of Civil Construction Engineering, 1995. Available at: <http://pcc2561.pcc.usp.br/textotecnicPCC16.pdf>. Accessed on: July 4, 2013.

AGUIAR, J.C. **Direito da Cidade**. Rio de Janeiro: Renovar, 1996.

AQUINO, N. A. **The construction of Belém-Brasilia and modernity in Tocantins**. Master's dissertation. Goiânia: Federal University of Goiás, 1996.

ARAÙJO, Suely Mara Vaz Guimaraes de. **The City Statute and the Environmental Issue**. Brasilia/DF: Chamber of Deputies - Legislative Consultancy, 2003. Available at: <www.camara.gov.br/internet/diretoria/conleg/Estudos/304366.pdf>. Accessed on: January 26, 2013.

BENEVOLO, L. **Storia dell' archittetura moderna**. Bari: Laterza, 1971.

BERNARDI, Jorge Luiz. **Social functions of the city**: concepts and instruments Master's dissertation. Curitiba: Pontificia Católica do Paranà, 2006.

BESSA, Kelly. **Contributions to the study of differentiation between urban centers**: the implications of the construction of Palmas for Porto Nacional in Tocantins. XIV ANPUR National Meeting, 2011. Rio de Janeiro: Proceedings of the XIV National Meeting of ANPUR, 2011.

BRAZIL. **Constitution of the Federative Republic of Brazil**. 1988. Available at: <http://www.planalto.gov.br/ccivil_03/constituicao/constituicaocompilado.htm>.

Accessed on: 26 Jan. 2013.

. **City statute**: guide for implementation by municipalities and citizens. Law n. 10.257, of July 10, 2001, which establishes general guidelines for urban policy. 2. ed. Brasilia: Câmara dos Deputados, Coordination of Publications, 2002.

CARVALHO FILHO, José dos Santos. **Commentaries on the City Statute**. Law No. 10.237, of July 10, 2001 and Provisional Measure No. 2220, of September 4, 2001. Rio de Janeiro: Lumen Jùris, 2005.

CARVALHO, Sonia Nahas de. **Condicionantes e Possibilidades Politicas do Planejamento urbano** In: VITTE, Claudete de Castro Silva; KEINERT, Tânia Margarete Mezzomo (orgs). Quality of life, urban planning and management: theoretical and methodological discussions. Rio de Janeiro: Bertrand Brasil, 2009.

CASTELLS, Manuel. **The urban question**. 4. ed. Rio de Janeiro: Paz e Terra, 2009.

CLARK, David. **Introduçao à Geografia Urbana**. 2. ed. Rio de Janeiro: Bertrand Brasil, 1991.

CORIOLANO, Germana Pires (eds.). State Report on the Evaluation of Master Plans in the State of Tocantins. In: SANTOS JUNIOR, O. A.; Montandon, D. T. (eds.). **Municipal master plans post-statute of the city**: critical assessment and perspectives. Rio de Janeiro: Letra Capital: Observatório das Cidades: IPPUR/UFRJ, 2011.

. **Palmas Participatory Master Plan**: an analysis of the application of urban planning tools to reduce socio-territorial inequalities. Master's dissertation. Palmas: UFT, 2011.

STATE OF TOCANTINS. **Official Gazette No. 3.990**. Year XXV, of October 24, 2013. Call notice.

FRANCISCO DE OLIVEIRA, A. Avaliação do Plano Diretor de Desenvolvimento Sustentàvel de Porto Nacional. In: SANTOS JUNIOR, O. A. e Montandon, D. T. (eds.) **Os pianos diretores municipais pós-estatudo da cidade**: balanço critico e perspectivas. Rio de Janeiro: Letra Capital: Observatório das Cidades: IPPUR/UFRJ, 2011.

FRANCO, Roberto Messias *et al.* **Main municipal environmental problems and prospects for solutions**. Municipios e meio ambiente, perspectivas para a municipalizaçao da gestão ambiental no Brasil. 1. ed. Sao Paulo: National Association of Municipalities and the Environment, 1999.

FREITAS, Henrique (orgs.). **The survey research method**. v. 35. n. 3. São Paulo: Revista de Administraçao, 2000.

GASPARINI, Diógenes. **The City Statute**. 1. ed. Sao Paulo: NDJ, 2002.

GIL, A. C. **Métodos e técnicas de pesquisa social**. 4 ed. Sao Paulo: Atlas, 1999.

GOITIA, F.C. **Breve História do Urbanismo**. Lisbon: Presença, 1992.

HAUSER, Philip Morris; SCHNORE, Leo F. **Estudo de Urbanização**. Sao Paulo: Pioneira, 1975.

IBGE. **PortoNacional population 2010**. Available at: <http://www.ibge.gov.br/cidadesat/topwindow.htm?1>. Accessed on: May 23, 2013.

IPHAN. **Historic Cities**. Available at: <http://portal.iphan.gov.br/portal/montarPaginaSecao.do?id=17776&retorno=paginaIph an>. Accessed on: 16 Apr. 2014.

LE CORBUSIER. **The Charter of Athens**. 1957. Available at: <http://www.scielo.br/pdf/spp/v14n4/9749.pdf>. Accessed on: July 4, 2013.

LEFEBVRE, Henri. **The Urban Revolution**. Belo Horizonte: UFMG, 2009.

LIRA, Elizeu R. **The genesis of Palmas** - Tocantins. Goiânia: Kelps, 2011.

. **The Genesis of Palmas**. TO. Master's dissertation. Presidente Prudente: UNESP, 1995.

MARICATO, Erminia. **Brasil, cidades alternativas para uma crise urbana**. 2. ed. Petrópolis, RJ: Vozes, 2001.

. **Housing and the City**. Coordination: Walderley Loconte. Sao Paulo: Atual, 1997.

MEIRELLES, Hely Lopes. **Brazilian Municipal Law**. 6. ed. Sao Paulo: Malheiros, 1993.

MENEZES, Luis Carlos Araujo; JANNUZZI, Paulo de Martino. Planning in Brazilian municipalities: A diagnosis of its institutionalization and degree of effectiveness In: VITTE, Claudete de Castro Silva; KEINERT, Tânia Margarete Mezzomo (eds.). **Quality of life, urban planning and management**: theoretical and methodological discussions. Rio de Janeiro: Bertrand Brasil, 2009.

MINAYO, M. C. de S. (org.). **Pesquisa Social**: teoria, método e criatividade. 22. ed. Rio de Janeiro: Vozes, 2003.

MUNFORD, L. **The city in history**. Belo Horizonte: Itatiaia, 1965.

OLIVEIRA, Adao Francisco. **EVALUATION AND TRAINING NETWORK FOR THE IMPLEMENTATION OF PARTICIPATORY MASTER PLANS PORTO NACIONAL SUSTAINABLE DEVELOPMENT MASTER PLAN** - PDDS-PN. Available   at:

<http://web.observatoriodasmetropoles.net/planosdiretores/produtos/to/TO_avalia%C3 %A7ao_pdp_portonacional_2009.pdf>. Accessed on: May 23, 2013.

OLIVEIRA, Aluisio Pires de; CARVALHO, Paulo Cesar Pires. **City Statute**: notes on Law 10.257, of July 10, 2001. Curitiba: Juruà, 2002.

OLIVEIRA, Isabel Cristina Eiras de. **Estatuto da cidade**: para compreender... Rio de Janeiro: IBAM/DUMA, 2001.

OLIVEIRA, Sebastiao de Souza. **Porto Nacional**: From Porto Real to the outskirts of Palmas - TO. Master's dissertation. Goiânia: UFG, 2009.

PEREIRA, Élson M. **Zoning and Social Housing.** Florianópolis, 2002.

PINTO, Lùcio Milhomem Cavalcante. **Luzimangues**: Social Processes and Urban Politics in the Genesis of a "New City". Master's dissertation. Palmas: UFT, 2012.

UNDP BRAZIL. United Nations Development Program. **Profile of Porto Nacional,** TO - Atlas of Human Development in Brazil 2013. Available at: <http://atlasbrasil.org.br/2013/perfil/porto-nacional_to#caracterizacao>. Accessed on: 17 Oct. 2013.

PORTO NACIONAL. **Autograph of the Municipal Organic Law**. Porto Nacional: Porto Nacional City Council, 1990.

. **Amendment to Organic Law No. 02/2011, of June 14, 2011**. Amends § 3 of art. 188 of the Organic Law of the Municipality of Porto Nacional, 2011.

. **Municipal Law No. 1.999/2009, of December 29, 2009**. Provides for the 2010/2013 Multi-Year Plan (PPA) for the Municipality of Porto Nacional, 2009.

. **Municipal Law No. 1883/2006, of December 22, 2006**. Provides for the Budgetary Guidelines of the Municipality of Porto Nacional for the 2007-2006 financial year.

. **Municipal Law No. 1884/2006, of December 22, 2006**. Provides for the updating of the Multiannual Plan - PPA, for the period 2006/2009 of the Municipality of Porto Nacional, 2006.

. **Municipal Law No. 2.132/2013, of December 5, 2013**. Provides for the Revitalization of the Sao Joao Stream in the Municipality of Porto Nacional.

. **Municipal Law No. 2015/2010, of October 8, 2010**. Provides for the creation of the Porto Nacional Municipal Environment Council, 2010.

. **PORTO NACIONAL SUSTAINABLE DEVELOPMENT MASTER PLAN** - PDDS-PN. Porto Nacional: Porto Nacional City Hall, 2006.

. City Hall. **Municipal Law No. 1887/2006, of December 22, 2006**. Establishes the Porto Nacional Municipal Environmental Policy Law, 2006.

PRIETO, Élisson Cesar. **The City Statute and the Environment**. 2006. Available at:<http://www.ibdu.org.br/imagens/OEstatutodaCidadeeoMeioAmbiente.pdf>.

Accessed on: May 19, 2013.

REZENDE, Vera F. Politica Urbana ou Politica Ambiental, da Constituiçao de 88 ao Estatuto da Cidade. In: RIBEIRO, Luiz C. de Q.; CARDOSO, Adauto L. (eds.). **Reforma Urbana e gestào democràtica**: promessas e desafios do Estatuto da Cidade. Rio de Janeiro: Revan/FASE, 2003.

RIBEIRO, H. Environmental impact study as a planning tool. In: PHILLIPI Jr.,A; ROMÉRIO. M.A.; BRUNA, G.C. (org.). **Curso de gestào ambiental**. Barueri: Manole, 2004.

RIBEIRO, Luiz Cesar de Queiroz. The City Statute and the Brazilian urban question. In: RIBEIRO, Luiz C. de Q.; CARDOSO, Adauto L. (eds.). **Urban reform and democratic management**: promises and challenges of the City Statute. Rio de Janeiro: Revan/FASE, 2003.

RODRIGUES, Ariete Moysés. **Boletim Paulista de Geografia** - "Perspectiva Critica" Environmental Problem = Political Agenda Space, territory, social classes. Sao Paulo: Association of Brazilian Geographers - AGB-SP, 2005. Available at: <http://www.usp.br/fau/depprojeto/labhab/biblioteca/textos/rodrigues_probl_ambiental. pdf>. Accessed on: 18 Apr. 2014.

RODRIGUES, Arlete Moysés. **Production and Consumption of and in Space - the urban environmental problem.** Editora Hucitec, 1998.

ROLNIK, Raquel. Urban Planning in the 1990s: new perspectives on old themes. In: Luis Ribeiro; Orlando Jùnior. **(Globalization, fragmentation and urban reform** - The future of Brazilian cities in the crisis. Rio de Janeiro: Civilizaçao Brasileira, 1994.

SANTOS JÙNIOR, Orlando Alves; MONTANDON, Daniel Todtmann (eds.). **Municipal Master Plans after the City Statute**: critical assessment and perspectives. Rio de Janeiro: Letra Capital: Observatório das Cidades: IPPUR/UFRJ, 2011.

SANTOS, Jocyléia Santana dos. Cenog in the discourse of its members. In: GIRALDIN, Odair (org.). **The historical (trans)formation of Tocantins.** 2. ed. Goiânia: UFG, 2004.

SANTOS, Milton. **Brazilian urbanization.** 5. ed. Sao Paulo: EDUSP, 2009.

SAULE JÙNIOR, Nelson. **New Perspectives on Brazilian Urban Law.** Constitutional Planning of Urban Policy. Application and Effectiveness of the Master Plan. Porto Alegre: Sergio Antonio Fabris Editor, 1997.

SEIFFERT, Mari Elizabete Bernardini. **Environmental Management**: instruments, spheres of action and environmental education. 2. ed. Sao Paulo: Atlas, 2011.

SEPLAN. Secretariat of Planning and Modernization of Public Management of the State of Tocantins. **Socioeconomic Profile of the Municipalities of Tocantins** - Porto Nacional. Palmas: Directorate of Research and Ecological-Economic Zoning, 2013.

SILVEIRA, Denise Tolfo; CÓRDOVA, Fernanda Peixoto. Scientific research. In: GERHARDT, Tatiana Engel; SILVEIRA, Denise Tolfo (eds.). **Research methods.** Porto Alegre: UFRGS,2009. Available at: <http://www.ufrgs.br/cursopgdr/downloadsSerie/derad005.pdf>. Accessed on: Aug. 22, 2013.

SOUZA, Marcelo L. de. **Changing the city**: a critical introduction to urban planning and management. 6. ed. Rio de Janeiro: Bertrand Brasil, 2010.

. **O desafio metropolitano**: um estudo sobre problemàtica sócio-ambiental nas metrópoles brasileiras. 2. ed. Rio de Janeiro: Bertrand Brasil, 2005.

SPÓSITO, Maria Encarnaçao Beltrao. **Capitalism and urbanization**. Sao Paulo: Contexto, 1998.

. The clash between environmental and social issues in urban areas. In: CARLOS, Ana Fani Alessandri; LEMOS, Amâlia Inês Geraiges (eds.). **Urban Dilemmas** - New approaches to the city. Sao Paulo: Contexto, 2003.

TORRES, Marcos Abreu. **City Statute**: its interface with the environment.

Goiânia: Revista Environmental Environmental, 2006. Available at :

<www.mpgo.mp.br/portalweb/hp/9/.../doutrina_estatuto_de_cidade.pdf>. Accessed on: September 25, 2013.

VELASQUES, A. B. A. **The 'last planned capital of the 20th century'**: the Palmas project (1989) and its modern condition. VIII SEMINARIO DOCOMOMO BRASIL, 2009. Rio de Janeiro: Proceedings of the 8th DOCOMOMO Brazil Seminar, 2009.

VITTE, Claudete de Castro Silva; KEINERT, Tânia Margarete Mezzomo (eds.). **Quality of life, urban planning and management**: theoretical and methodological discussions. Rio de Janeiro: Bertrand Brasil, 2009.

ANNEXES